Happy Pills in America

Happy Pills in America

From Miltown to Prozac

DAVID HERZBERG

The Johns Hopkins University Press

Baltimore

Johns Hopkins Paperback edition, 2010
9 8 7 6 5 4 3 2 1

The Johns Hopkins University Press
2715 North Charles Street
Baltimore, Maryland 21218-4363
www.press.jhu.edu

The Library of Congress has catalogued the hardcover edition of this book as follows:

Herzberg, David L. (David Lowell)
   Happy pills in America : from Miltown to Prozac / by David Herzberg.
      p. ; cm.
   Includes bibliographical references and index.
   ISBN-13: 978-0-8018-9030-7 (hardcover : alk. paper)
   ISBN-10: 0-8018-9030-6 (hardcover : alk. paper)
   1. Psychotropic drugs—United States—History. 2. Psychotropic drugs—Social aspects—
United States. I. Title.
   [DNLM: 1. Psychotropic Drugs—history—United States. 2. Psychotropic Drugs—
therapeutic use—United States. 3. Consumer Satisfaction—United States. 4. Culture—
United States. 5. Drug Industry—trends—United States. 6. History, 20th Century—
United States. QV 77.2 H582h 2008]
   RM315.H52 2008
   362.290973—dc22   2008007645

A catalog record for this book is available from the British Library.

ISBN 13: 978-0-8018-9814-3
ISBN 10: 0-8018-9814-5

*Special discounts are available for bulk purchases of this book. For more information,
please contact Special Sales at 410-516-6936 or specialsales@press.jhu.edu.*

For Alvar

CONTENTS

Acknowledgments   ix

Introduction. Medicine, Commerce, and Culture   1

1   Blockbuster Drugs in the Age of Anxiety   15

2   Listening to Miltown   47

3   Wonder Drugs and Drug Wars   83

4   The Valium Panic   122

5   Prozac and the Incorporation of the Brain   150

Conclusion. Better Living through Chemistry?   192

Appendix A. Medications Mentioned   205
Appendix B. Prescriptions for Psychiatric Drugs,
    1955–2005   207
Notes   209
Index   275

ACKNOWLEDGMENTS

A great many people helped make this book possible. My thanks begin where the project did: with Paul Boyer, a generous mentor who advised students to "follow their bliss" and look for history in unexpected places, even if it took them away from his own areas of expertise (a difficult feat given his many interests). Judy Leavitt's warmth and excellent skepticism helped orient me in the history of medicine and reminded me that people, not "culture," make history. Thanks too to those who read early drafts of individual chapters, including Hiroshi Kitamura, Chris Wells, Gwen Walker, and David Musto. I was very fortunate to have funding in the project's early stages from a Jacob K. Javits Fellowship, and I owe thanks for research support from the American Institute for the History of Pharmacy, the University of Wisconsin History Department, and the Wisconsin Alumni Research Foundation.

The University at Buffalo has been an excellent place to research and revise. Colleagues have been supportive, and department chairs work hard to protect time for scholarship. The university's Julian Park Fund also helped defray publication costs. Susan Cahn provided moral support and invaluable comments at a late stage of revisions. Like most historians I owe many thanks also to librarians, including those at Lockwood and the Health Sciences Library at SUNY Buffalo, the Buffalo Public Library, the National Library of Medicine, and the New York Academy of Medicine (the last being one of the only U.S. libraries that did not cut out drug advertisements when they bound medical journals— a circumstance that at least some of their staff must have regretted when photocopying the gazillionth ad for me). I also have benefited from the suggestions of anonymous reviewers for *American Quarterly*, the *Bulletin of the History of Medicine*, and the Johns Hopkins University Press. I am lucky to have had the guidance and support of my editor at Hopkins, Jacqueline Wehmueller, who has been generous with her time and insight. Finally, Verispan LLC generously donated prescription survey reports, for which I am grateful.

Some people have helped so much that very little can be said about them without writing another whole book. Nan Enstad continually inspired me to see and work toward the best possibilities in the project. Nancy Martina and her allies gave crucial guidance in the last stages of revision. Don and Vickie Herzberg have given unflagging support throughout; this work would have been impossible from the start without them. And finally, three people who have never known me when I was not working on this book: Rex, Leo, and Erin. You are my happy pills.

Happy Pills in America

Medicine, Commerce, and Culture

IN 2001 THE GlaxoSmithKline Group of Companies announced good news for the "10 million people who live with excessive uncontrollable worry, anxiety, tension, irritability, restlessness and sleep disturbances": relief, in the form of Paxil, an antidepressant, was at hand. Advertising in *Newsweek,* the pharmaceutical giant advised "chronic anxiety" sufferers to "talk to your doctor about non-habit-forming *Paxil* today. So you can see someone you haven't seen in a while . . . Yourself." The main image showed a worried-looking woman striding purposefully down the street. The tag-line promised, "Your life is waiting."[1]

Few Americans at the turn of the twenty-first century would have been surprised by this ad or its claims. The notion that pills could restore selfhood had become commonplace, pervading popular as well as medical culture. Advertisers were not the only ones spreading the gospel of neurochemistry, the idea that consciousness and selfhood were manifestations of biology, and therefore both knowable and controllable. According to this gospel, brains were like other parts of the body, capable of malfunctioning and leaving their owners painfully "not themselves." Luckily, advances in medical science promised to fix the mechanical breakdowns ("correct the chemical imbalance believed to cause the disorder," as the Paxil ad put it) and bring people back to health—to "themselves." For people suffering from depression and its very real miseries such promises meant a great deal, especially if (as was likely) they personally knew someone who had been helped by Paxil or one of its cousin drugs. The daily experiences of millions of Americans helped make possible the claim, once the province of LSD proselytizers like Timothy Leary, that taking a medicine—a drug—could lead to a more authentic self.

Nor would early-twenty-first-century Americans have been shocked to see a powerful prescription medicine like Paxil advertised in a popular

1

Does the woman make the syndrome? *Source:* Paxil ad, *Newsweek*, September 17, 2001.

magazine like *Newsweek.* For most of the twentieth century, "ethical" pharmaceutical companies had advertised only to physicians in medical journals and other professional forums. By 2001, however, "direct-to-consumer" ads in magazines and on television had become familiar fare. These advertisements helped entrench the notion of patients as consumers who purchased pharmaceutical goods through their doctors. In combination with breathless reporting about drug sales, the expensive ads also served as continual reminders that antidepressants like Paxil were "blockbuster" drugs, earning enormous profits for their manufacturers. In this context, the claim to be able to deliver "yourself" took on the accents of the consumer culture, implying a lifestyle choice as well as a medical therapy.

The most familiarly comforting part of the advertisement, however, making the other claims seem safe and unobjectionable, might have been the image. The woman is white and well-dressed; her neck and the uncovered portion of her arm look well-toned; she walks among blurry figures whose coats and ties are nevertheless visible. She therefore represented what Americans were likely to have already known or suspected about antidepressants: of the tens of millions who used them, far more were white than nonwhite, well-off than poor, and (by a two-to-one margin) women than men. Public discussion of the drugs almost always appeared in white middle-class media, often women's magazines, and almost universally related to the drugs' meaning for white middle-class culture. The contrast with "narcotics," associated with nonwhite and marginal groups, would have been obvious. It was, therefore, the woman in the ad who turned the constellation of otherwise fairly common symptoms—fatigue, sleep problems, worry, restlessness, muscle tension, anxiety, and irritability—into "Generalized Anxiety Disorder," a legitimate medical problem calling for "non-habit-forming" mood-enhancing drugs.

The Paxil ad might seem to be a recent phenomenon, but its roots stretch back to the origins of modern medicines in post–World War II America. The first blockbuster psychiatric medicine, the minor tranquilizer Miltown, was introduced in the 1950s, and use of the drugs reached a peak as early as 1973, when almost one hundred million prescriptions were written for Miltown's successor, Valium, alone. Long before direct-to-consumer advertisements like Paxil's, drug marketing campaigns supposedly limited to doctors' eyes were designed to "escape" the medical world and spread into popular consciousness—informal campaigns that suffered few if any of the restrictions imposed on advertisements in later decades. And throughout these years, tranquilizers and antidepressants served as icons in the inner lives of the white middle classes, defined in part through contrast with illegal "street" drugs, and praised or vilified for what they revealed about the nature and fate of white middle-class culture.

After psychiatrist Peter Kramer reached the bestseller lists with *Listening to Prozac* in the 1990s, it became fashionable to wonder about the significance of antidepressants for concepts of consciousness, identity, and selfhood—what Kramer called "the message in the bottle." I argue that the pill bottle holds other kinds of messages, too, and not just about neurology and personal well-being. The meteoric rise of tranquilizers and antidepressants signified broader developments in American society after World War II: the commercialization of medicine and science, the embrace of psychology and self-fulfillment as a political language, intensified campaigns to police social groups through drug regulation, and social movements organized in part around new concepts of identity.

These broad social transformations revise our understandings of postwar American history in at least three significant ways. First, they open new windows onto the expanding postwar consumer culture. Much has been written about tranquilizers' and antidepressants' impact on the medical professions—how biological psychiatrists brought an end to Freudian dominance, for example, or how pharmaceutical companies influenced the diagnostic categories of mental illness.[2] I look in a different direction, examining the commercial development and popular marketing of blockbuster medicines after World War II. This story shows how the expanding consumer culture not only enveloped individual buyers but, by making consumers out of doctors and patients, reconfigured social institutions such as the medical system. Like suburban houses, new cars, and washing machines, medicine became part of a new consumerist "American dream" that reconfigured conceptions of what a good middle-class life—what happiness itself—ought to be like.[3]

Second, these stories bring the study of social movements into dialogue with other broad developments in postwar history. Postwar American histories of subjects like technology and consumerism, and histories of social movements like feminism, have not always had much to say to each other; they focus on different dynamics, different approaches, and different actors. My study highlights the ways in which these stories are important to one another: social movements have both shaped and been shaped by their participation in other major dynamics such as the spread of consumerism, the rise of therapeutic culture, the diffusion of new technologies, and antidrug wars. Thus, for example, feminists were key actors in fashioning the meanings and uses of blockbuster tranquilizers; these efforts, in turn, affected feminists' own political interpretation of identity and self-fulfillment.

Third, I take a new perspective on the twentieth century's "wars against drugs." Because tranquilizers and antidepressants are prescription medicines, they have not typically been examined as part of the history of their feared cousins, the "street" drugs. But they are hardly alone among mind-altering drugs in having medical value; most street drugs made their debuts as wondrous new tools of medicine. Indeed, it is just such tensions as those between the ever-changing categories of "medicines" and "drugs" that make tranquilizers and antidepressants such valuable subjects of inquiry. Focusing largely on illegal drugs, historians have shown how antidrug campaigns have served the public health but also have been used by state authorities to police nonwhite, poor, or otherwise marginal populations both domestically and globally.[4] My study of tranquilizers and antidepressants suggests that this history of drug wars is inextricably entwined with the history of wonder drugs—that, in fact, each depended on the other for meaning and power. The binary opposition of street and medicine cabinet is unstable and has changed over time, opening unexpected opportunities to construct or challenge the social hierarchies built into American drug policies.

As these examples suggest, this book is as much about people as it is about drugs. The importance of Miltown, Valium, and Prozac emerged not just from their pharmacology but from the actions of many different groups of Americans: physicians and patients, yes, but also corporate executives, advertisers, journalists, political activists, and many others. Each of these groups struggled over the place of "mind drugs" in medical practice, in consumer society, and in political cultures as they sought to take advantage of personal, professional, or political opportunities. It was their actions, and not the drugs themselves, that produced the dramatic transformations in American medicine, commerce, and culture that have reshaped the very meaning of identity since World War II.

Physicians and patients looking for ways to relieve genuine suffering were central to the rise of tranquilizers and antidepressants, of course. Doctors embraced the drugs eagerly, prescribing them by the millions, while patients willingly accepted or even demanded them. Within a decade of the introduction of these drugs, prescribing a minor tranquilizer had become one of the most common therapeutic actions in American medicine. By the early 1970s, Valium alone had become the most prescribed medicine in America and in the rest of the world (see appendix B). The availability of these drugs helped cement medicine's expanded postwar commitment to psychological health, providing ordinary physicians and their patients new ways to combat anxiety, tension, depression, and the ill-defined troubles of the stomach, skin, and heart that often accompanied them. Patients have always sought counsel from physicians on such matters, but tranquilizers and antidepressants expanded and formalized this aspect of medical practice and helped reshape doctor-patient relationships.[5]

Key facilitators in this formalized relationship were drug marketers, who took on new importance as they underwrote the psychotropic phenomenon with a profusion of clever advertisements and public-relations ploys. Pharmaceutical companies had always been commercial enterprises, of course, even the "ethical" or prescription drug firms that advertised only to physicians. But the postwar boom created what seemed like a wholly new drug industry. Drug companies now routinely sold antibiotics and the era's other new wonder drugs as brand-name products, marketing them with unprecedented intensity not just to physicians but also to the general public through a variety of creative means. Tranquilizers were among the most profitable of the new brand-name drugs —the leading products of an industry that was itself one of the most profitable sectors of the booming postwar economy. Commercial dynamics, in other words, shaped American notions of medicine from the earliest days of modern miracle drugs, beginning with Miltown, the first true blockbuster.[6]

In some ways, drug advertisers' agendas dovetailed with those of doctors and patients. Advertisers described the tranquilizers and antidepressants as truly revolutionary medical advances, helping to anchor them in the therapeutic arsenal as "medicines" rather than as "narcotics" or "dope." In doing so they added their voices to psychiatrists' calls to recognize the medical significance of psychological suffering. At the same time, however, advertisers had their own take on the meaning of psychological illness, one designed to cast the widest possible net for potential customers. Their campaigns thus tended to conflate psychological illness with the familiar daily problems that populated the cultural landscape of consumerism. These problems—marital discord, frustration with traffic, housekeeping woes, an inability to "fit in"—were undoubtedly real sources of misery, but they had not always been interpreted as illnesses.

Drug advertisers targeted physicians but also designed their appeals to escape into popular consciousness. Their messages enjoyed a wide, almost pervasive, circulation in medical and popular circles, playing a central role in establishing the nature and meaning of the new medicines. The ubiquity of the campaigns, and the well-reported profits of the drug companies that directed them, spread an idea that subtly altered the medical message: tranquilizers and antidepressants were consumer goods. And like other goods, they were available to bring comfort and convenience to consumers' psychic and emotional lives. This idea had particular resonance in the postwar era, a time when Americans were increasingly taught to see consumer spending as a civic duty and a badge of political freedom, not just a quest to fill personal needs.[7]

Important as they were, however, advertisers were not the only voices shaping the psychotropic phenomenon. Their success in projecting the new wonder drugs into public awareness invited others to participate too, people who saw opportunities to use the ubiquitous new drugs to pursue their own cultural and political agendas. The result was controversy—a dynamic public dialogue about the meaning of wonder drugs for the mind, deeply conditioned but never fully dictated by the forces of commercial medicine.

Some of these controversies were relatively lighthearted. The "happy pill" Miltown, for example, drew the often-disapproving attention of such luminaries as Margaret Mead and the Pope, and it was briefly the talk of Hollywood, where comedian Milton "Miltown" Berle claimed that it was indispensable. At the same time, a wide range of cultural commentators and social activists seized on the pills' ubiquity to dramatize political arguments or even to organize new kinds of activism. This activism was largely based in white and middle-class culture, perhaps in part because tranquilizers (and, later, antidepressants) were treatments for problems similar to the "nervous illnesses" that had long been considered diseases of affluence. As a result, it was easy to see the new wonder drugs as having special meaning in the ongoing cultural drama of the inner lives of America's white-collar classes, just as narcotics loomed large in the symbolic life of less privileged groups.

From Miltown to Prozac, critics have pointed to psychiatric medicines as both the symbol and the substance of deeply gendered crises in white middle-class culture. The first of these crises surfaced in the 1950s, amid a broad effort to inscribe new gender roles claimed to represent a return to tradition. As white-collar men struggled to adapt to their new positions as family patriarchs, consumers-in-chief, and cogs in the corporate world, cultural critics revived old fears of "race suicide" in new ways. They held up Miltown as the ultimate consumer good, an icon of the soft conformity that threatened to destroy the nation's "best men" by comforting them into submission. In the 1960s and 1970s, one segment of the feminist movement adapted this argument as a political critique of middle-class women's social position. Librium and later Valium, prescribed for women at twice the rates of men, became symbols of how society limited affluent women's self-fulfillment and kept them safely ensconced in the home. These criticisms only became sharper during a panic over Valium addiction in the 1970s. In later decades Prozac became a blockbuster because the drug's supporters successfully portrayed it as a nonaddictive, commercially available self-enhancement technology that helped women become "supermoms" with careers and loving families—the perfect solution to the Valium crisis.

The medicines whose stories I follow in this study are blockbusters: minor tranquilizers like Miltown and Valium and antidepressants like Prozac and Paxil. These were the most widely used, the most profitable, and the most culturally controversial of the modern psychiatric drugs. Their importance on these multiple levels helps explain why I focus on minor tranquilizers and antidepressants rather than on what was arguably the more medically significant major tranquilizers or antipsychotics.

The *major tranquilizer* Thorazine, introduced in the early 1950s, was the first miracle drug for the mind—indeed, it was the medicine that established the miracle-drug template that Miltown and its successors adapted to such great effect. (Miltown's marketers were happy to secure the name minor tranquilizer to suggest an affinity between their drug and Thorazine, despite the almost complete lack of pharmacological similarity between the two.) Though no one knew why, Thorazine appeared to have some specific effect on the mechanisms of psychosis. By improving psychotics' ability to live independently, Thorazine and its successors (later known as *antipsychotics*) helped facilitate psychiatry's shift away from asylums to community-based care. Nonetheless, the antipsychotics never reached the commercial or cultural status of the more widely used tranquilizers and antidepressants.

Miltown, Valium, and the *minor tranquilizers* (later known as *anti-anxiety agents*), meanwhile, were relatively mild muscle relaxants and sedatives prescribed to ease tension or anxiety. They were called minor because their effects were relatively mild, and they appeared to have little or no impact on people with serious mental illnesses. The most important minor tranquilizers were Miltown and its chemically identical competitor Equanil, both introduced in the mid-1950s, and their successors Librium and Valium, which became available in the early 1960s.

Antidepressants such as Prozac and its precursors—initially referred to as *psychic energizers*—were used to elevate mood, reduce suicidal impulses, and counteract a range of other depressive symptoms such as insomnia, anxiety, and disturbance of appetite. The best known antidepressants were Prozac and its competitors Paxil and Celexa, first available in the late 1980s, but a wide range of earlier antidepressants had been in (much lesser) use since the late 1950s.

Advocates of minor tranquilizers and antidepressants touted them as a major advance over earlier drugs. *Barbiturate sedatives,* for example, had been available since the early 1900s, but like alcohol and the opiates, these drugs stupefied as they calmed; users acted drunk, with impaired thinking and physical control. *Amphetamine stimulants* were first made available in the 1930s, but like cocaine they could elevate energy above normal or healthy levels into something approaching mania.[8] Both barbiturates and amphetamines tended to give users an immediate, intense high that fostered dependence or addiction. According to their supporters, minor tranquilizers and antidepressants solved these problems: their effects were predictable, they produced no high or addiction, and they improved rather than impaired social functioning. These distinctions seemed more readily apparent for the antidepressants, which took two weeks to produce their effects and seemed to act directly on some underlying mechanism of depression, than for the minor tranquilizers, which even to some psychopharmacologists seemed little more than improved versions of existing sedatives.

That Thorazine was so quickly overshadowed by the far less powerful Miltown and its successors underlines the importance of economic and cultural factors in shaping the meanings of the new drugs. Because their use was limited to the relatively small population of the seriously mentally ill, Thorazine and later antipsychotics never attained the commercial or cultural importance of Miltown or Valium. Early antidepressants languished in a similar condition. They were genuinely exciting new treatments for an intractable illness, but physicians rarely used them, and they garnered little public attention until the late 1980s, when depression had gained prominence, the tranquilizers' star had dimmed, and easier-to-use new antidepressants like Prozac had emerged. Indeed, for much of the 1960s and 1970s, advertisers for antipsychotics and antidepressants often portrayed their drugs as slightly more powerful versions of the minor tranquilizers, helpful for "everyday" patients with common troubles. At the same time, manufacturers of older drugs like barbiturates and amphetamines, and even over-the-counter sleep aids and caffeine pills, marketed their products as tranquilizers or antidepressants.

To emphasize such economic and cultural factors is not to imply that the sufferings of the anxious or depressed, and the relief provided for many by tranquilizers and antidepressants, have simply been social constructions with no basis in reality. In fact, the cultural dynamics of this story are so powerful precisely because real suffering and real relief were at stake. It was just such matters of psychological well-being that made the nervous illnesses, and the medicines used to treat them, so potent as both symbol and substance in broader debates over social welfare. The important experiences of suffering and relief meant many different things to different people. Some of these meanings came to be championed by one or another group (advertisers, feminists, drug reformers, etc.), gaining cultural visibility and influence—however far afield they may have strayed from the real experiences of most drug takers.

This, in short, is a cultural history, tracing how different groups responded to the medical, commercial, and cultural opportunities presented by psychotropic medicines in the post–World War II era. It explores how these responses shaped understandings and uses of the drugs. It also looks in the other direction, arguing that a political engagement with wonder drugs changed not just the history of the drugs but of those who engaged them, too. Choosing Valium as an icon of sexism had consequences for the feminists who pursued the tactic, just as advertising medicines as the ultimate consumer good pushed commercial culture in new directions.

This book is not an attempt to reconstruct the experience of having a mental illness or of taking a medicine, or, for that matter, of prescribing one. Nor do I resolve the controversies that have long surrounded the drugs: how effective and safe they are, for example, or whether they ought to be prescribed so often. Medical and cultural assumptions about these and other issues have changed over time, sometimes dramatically, and the ferment shows little signs of ending. For the cultural historian, what fascinates about these beliefs is not their truth or falsity but how they came to be persuasive, to whom, and with what effects. Thus this study neither praises tranquilizers and antidepressants as technological saviors nor damns them as narcotic enslavers, but rather seeks to understand why *these* stories, and not others, have successfully defined the drugs in American culture. It is the story of how people transformed the worlds of medicine, commerce, and culture through their efforts to make sense of—and take advantage of—an important new technology.

Chapter 1 examines the origins of modern psychotropic medicines in the 1950s and 1960s. The emergence of Miltown and other wonder drugs has typically been thought of as a tale of scientific breakthroughs, and the years after World War II as a golden age of discovery and medical advance. The idea of medicines as intensely advertised consumer goods, and as blockbuster earners for companies and stockholders, seems of far more recent vintage. But wonder drugs were fundamentally commercial and cultural creations from the very beginning. Long before Prozac and direct-to-consumer advertising, Miltown and other modern medicines were developed, marketed, and distributed through commercial and consumerist systems that transformed the institution of medicine and put an indelible stamp on the very concept of medicine in postwar America. This chapter looks at how advertisers, physicians, and patients worked together to construct this system, each for their own reasons, with important and often surprising consequences.

Promoted by advertisers and demanded by physicians and patients, the tranquilizers almost immediately joined the ranks of America's most-prescribed drugs, used by millions. Chapter 2 examines how the drugs also became a pop-culture phenomenon, propelled by a range of critics from medical journals to the front pages and gossip columns of popular media. Although the tranquilizers would later become known as "women's drugs," they first gained fame in the 1950s for their supposed impact on men. Amid the broad cultural campaigns to reinstate "traditional" gender roles in postwar America, critics denounced Miltown—widely depicted as a medicine for the affluent—as a threat to white-collar masculinity. Advertisers responded with new campaigns promising that their drugs would restore, not undermine, manly character, helping to stamp the emerging biological psychiatry with the gendered logic of the 1950s.

In the 1960s, early second-wave feminists like Betty Friedan adapted anti-Miltown arguments for their own purposes. They held up tranquilizers like Librium and Valium as archetypal examples of stifling sexist responses to the dissatisfactions of well-educated housewives. Both sets of critics updated old fears of "race suicide" for postwar America, linking tranquilizers and other drugs with basic debates over middle-class identity and altering both politics and medicine in consequential ways.

Chapter 3 examines how a small but growing number of physicians, drug researchers, and activists destabilized drug-war verities in the 1960s by spreading word that tranquilizers (and other "mind drugs") could be addictive. During the twentieth century's campaigns against drugs, addiction had largely been associated with marginal populations and their narcotics. Miltown, Librium, and Valium were legal prescription medicines, scientific wonders, and highly popular consumer goods—staples of the respectable classes. The campaign to prove that such medicines might be addictive just like "dope" stimulated a decade-long controversy that pitted traditional antidrug allies against each other, and ultimately altered the war against drugs by successfully incorporating the tranquilizers as controlled substances. This attenuated victory—a more lenient category was created for the minor tranquilizers, specifically designed to limit any stigma—illustrated how economic and cultural agendas not only shaped American drug policy but also reconfigured basic concepts such as addiction itself.

Even though many physicians and drug regulators still viewed it favorably, Valium became the target of one of America's periodic drug panics in the late 1970s. Popular media aired fears of an epidemic of addiction to the drug, Congress held repeated inquiries, and famous figures like former first lady Betty Ford and television producer Barbara Gordon told stories of their addiction in best-selling memoirs. By the early 1980s, use of the onetime wonder drug was in steep decline, its reputation sullied by suspicion that it was just a form of legal dope. Chapter 4 analyzes the cultural politics of the Valium panic in the context of other episodes of American drug hysteria. While many different groups contributed to the panic, the Valium affair was largely fashioned by the white middle-class wing of the diverse second-wave feminist movement. They adapted established antidrug rhetoric to sensationalize Valium addiction among affluent white women as a central symbol of sexism and its consequences. And they held up liberation from "mother's little helper" as an archetypal story of emancipation through feminism. This was a remarkable campaign, attracting new audiences for feminist political messages and challenging the punitive logic of the twentieth-century's war against drugs. It was also, however, limited in important ways by the class and race dynamics of American drug politics.

Chapter 5 examines the rise of depression in the 1970s as America's most common emotional illness, and the emergence of Prozac and a new generation of antidepressants as the public face of psychopharmacology in the 1990s. Prozac's popularizers promoted the new wonder drug as a resolution to the Valium crisis. It was, they said, nonaddictive, it had few side effects, and as an energizer it liberated instead of pacified the women who (still disproportionately) took it. Boosters claimed that these valuable qualities were the result of a revolutionary scientific advance: the discovery of a selective medicine that targeted a single brain neurotransmitter, serotonin. Prozac and serotonin became pop-culture icons, symbols of a biological revolution that defined happiness, creativity, love, and identity itself as manifestations of neurochemistry. The Prozac craze helped place a consumerist stamp on this biological revolution, favoring relatively simple models of the brain that made it possible to imagine (and advertise) new blockbuster drugs delivering the good life directly to the psyches of America's comfortable classes. By linking Prozac to free consumer choice rather than sexist medical systems, these stories also resolved broader political arguments raised during the Valium crisis. In an era of organized resistance to feminism and of reinvigorated drug wars, this seemingly apolitical story of technological triumph appealed in many quarters. But in popularizing Prozac, the drug's boosters also opened it to criticism as one of many new technologies for remaking the self that have marched under the banner of science while being implemented under the system of commerce.

Today, many biotechnologies—cloning, drugs, genetic engineering, and so forth—promise to open people's bodies and consciousness to capitalism's "creative destruction" as never before.[9] To navigate this profusion of new possibilities, it is necessary to understand their origins. In this book I argue that the opportunities and dangers presented by a signal new technology were not inherent in the technology itself but emerged from the way in which people seized on new discoveries to pursue old agendas: to ease suffering, to earn profits, and to affirm or challenge the politics of identity. It was these efforts that led the consumer culture to encompass new territory in human consciousness; that altered concepts of identity so crucial to modern American political strivings; and that reconfigured major political institutions like the war against drugs. This was ultimately a story driven by people, not by drugs—and it still is today. Even the mammoths of the medical-industrial complex that emerged after 1945 shaped history only in dynamic interaction with ordinary people in pursuit of betterment of their lives.

CHAPTER ONE

Blockbuster Drugs in the Age of Anxiety

BY ALL ACCOUNTS, the 1950s and 1960s were the heyday of Freud in American medicine and culture. Psychoanalysts chaired the vast majority of prestigious medical school psychiatry departments, where most students learned from psychoanalytically oriented textbooks and curricula. The American Psychiatric Association too was dominated by analysts, and in 1952 the organization joined with the Association of American Medical Colleges in advocating psychoanalytic training for all psychiatrists. Freudians' influence extended beyond the halls of medicine as well, as a simplified version of psychoanalytic ideas seeped into the culture at large, placing its stamp on everything from movies to Dr. Benjamin Spock's best-selling child-rearing bible, *Common Sense Book of Baby and Child Care*. Freud was a name to reckon with, and the figure of the analyst, bearded and all-knowing, was the popular face of psychiatry.[1]

Freud's grip on American medical and popular imagination would seem to have boded poorly for drug therapy. Freudian or "psychodynamic" psychiatry taught that people's experiences, not the physical state of their brains, caused mental illness—that twisted thoughts came from twisted experiences, not twisted molecules. Direct "repairs" of brain matter through surgery, electricity, or drugs thus seemed irrelevant or even harmful. Indeed, Freud's postwar followers have been criticized for using their enormous medical and cultural influence to delay for decades the triumph of biological psychiatry and its revolutionary drugs.

In reality, however, drug therapy flourished during the "age of Freud." Literally tens of millions of Americans took antipsychotics, antidepressants, and, most of all, antianxiety agents like Miltown in the 1950s and 1960s. Drug takers, in fact, vastly outnumbered their peers in psychoanalysis, and quite probably surpassed the number of troubled souls in any other kind of formal therapy, "talk" or otherwise. As historian Jonathan Metzl has noted, the "age of Prozac" actually started decades before Prozac was discovered and does not appear to have ignited the titanic battle with Freudians that many have assumed.[2] Indeed, as Miltown became a celebrity to rival Freud, the two often appeared in the same issue of popular magazines like *Newsweek* with nary an editorial eyebrow raised.

How did Miltown and competitor drugs succeed so well in such a seemingly inhospitable climate? The answer has a great deal to do with the commercialization of American medicine after World War II. The famed miracle cures of the 1950s—antibiotics, the polio vaccine, and so forth—were not just therapeutic tools. They were among the nation's most successful commercial goods, manufactured and sold for profit by one of the signature industries of the postwar boom economy. Intense marketing spread word of new drugs like Miltown, the first true block-buster prescription medicine, to physicians and popular audiences alike. Aided by helpful advice from advertisers, ordinary physicians eagerly embraced new tools in what they still saw as an essentially Freudian quest to treat anxiety. Indeed, it was their very acceptance of Freud's theories about anxiety that helped persuade them to reach for the prescription pad. Public-relations campaigns in popular media, meanwhile, portrayed "happy pills" as the latest new technology for achieving the good life and encouraged potential pill takers to behave like consumers and actively seek new pharmaceutical marvels from their doctors. Countless Americans followed this advice, but not always as physicians or advertisers hoped they would.

Ultimately, drug companies, physicians, and patients worked together, each for their own reasons, to create what we might call a medical-industrial complex with great economic, social, and cultural power. This system marshaled enormous financial and marketing resources from one of the most formidable segments of the economy. It conquered new territory for the consumer culture by transforming doctors and patients into consumers, and by opening up the psyche as the newest frontier for capitalism's "creative destruction." It had the potential to change millions of lives by providing people with new drugs to help them combat emotional and psychological pain. And its mixture of consumerist and therapeutic promises conjured new visions of the good life that could inspire new efforts to pursue personal and political well-being.

Wonder Drugs and Blockbuster Drugs in the Golden Age of American Medicine

Miltown was just one among many wonder drugs that appeared in the fifteen or so years after World War II, including antibiotics, vaccines like the Salk polio vaccine, hormones like steroids and the birth control pill, and antipsychotics like Thorazine. These new drugs transformed medical practice in America, giving physicians truly curative tools for a wide range of illnesses. It is almost impossible to conceive of modern therapeutics without them.

The emergence of these miraculous new drugs bears all the hallmarks of a classic story of scientific discovery. It is a tale full of individual brilliance, inexplicable hunches, and farsighted pioneers battling the conventional wisdom of their day—the very stuff of heroic science. But the story also reveals something else: intensifying entanglement of the prescription drug industry with the booming commercial economy. Heroic researchers may have discovered the new chemicals, but the path to making those chemicals available as medicines—and to spreading the news to physicians—increasingly led through the marketing department rather than the laboratory. And as we shall see with Miltown, in many cases the research itself originated in brilliant hunches of the commercial as much as the scientific variety. The wonder drugs of the 1950s, in other words, heralded a revolution in medical economics as well as in therapeutics. Although often overshadowed by the antibiotics, tranquilizers and antidepressants were crucial at both levels of this medical revolution.[3]

The first and probably most important of the new mind drugs was chlorpromazine, better known by its brand name Thorazine. In the late 1940s, French surgeon Henri Laborit, grappling with the problem of patients going into shock while under the knife, turned to a recently discovered class of drugs known as antihistamines. These drugs inhibited the body's allergic responses but were known to have an unwanted side effect: sedation. Laborit found that these drugs kept patients calm during surgery and also mitigated shock reactions. His success in exploiting these side effects provoked renewed interest in a family of antihistamines known as the phenothiazines, which appeared to have unusually strong calming effects. By late 1950 Frenchman Paul Charpentier, discoverer of the phenothiazines, had synthesized a molecule that was virtually *all* side effect: chlorpromazine. It was hardly antihistaminic at all, but was a powerful calming agent. According to an impressed Laborit, the drug "provokes not any loss in consciousness, not any change in the patient's mentality but a slight tendency to sleep and above all 'disinterest' for all that goes on around him."[4]

Perceptive observers like Laborit immediately recognized chlorpromazine's potential value for the treatment of mental illness, and the drug soon began to make its way into the psychiatric armamentarium throughout Europe. In America, however, where Freudians ran most mental hospitals, few drug companies were eager to bet on a drug to treat psychotics. Smith Kline and French, the only U.S. firm interested in chlorpromazine, ignored the drug's antipsychotic properties in favor of a useful side effect; in 1954 they marketed it, under the name Thorazine, as an antinausea drug. To those asylum psychiatrists who kept up on such matters, and who were eager to experiment with cutting-edge therapies, however, Thorazine immediately offered other prospects. When treated with the drug, many psychotic patients—even the most chronic and intractable—quickly regained contact with reality. They became rational, calm, and accessible to treatment. The "disinterest" that Laborit had noticed turned out to be the hallmark of a new kind of relaxant, at first dubbed tranquilizers to distinguish them from sleep-inducing sedatives such as the barbiturates, ether, or chloral hydrate. Later they would be known as antipsychotics.

Although Freudians still ruled from their influential professional and academic posts, Thorazine spread with great rapidity in American institutional practice. In the first eight months after it became available, it was prescribed over two million times. Supportive observers detected and applauded an easing of the raucous, "bedlam" quality of mental hospital wards. Many psychotics could now be treated with drugs outside the asylum—an appealing circumstance for a psychiatry increasingly focused on community therapy. Partly as a result of this advance,

American asylum populations declined for the first time since records had been kept.

The tranquilizers were not universally loved, however. They did not help every patient, and they did not cure. Patients who stopped drug therapy typically relapsed. To a significant portion of observers, especially committed Freudians, this suggested that the drugs merely covered up, rather than healed, a person's illness, much like straightjackets and the other physical treatments that talk therapy had supposedly replaced with more humane techniques. Such a possibility provoked searing critiques about mind control from radical therapists Thomas Szasz and Joost Meerlo, later to win notoriety as key figures in the 1960s' antipsychiatry movement.[5] Their criticism was given further strength by the tranquilizers' troubling array of potential side effects, which ranged from the unpleasant (jaundice) to the disfiguring (Parkinson-like jerks and trembling) to the potentially fatal (one variety appeared in rare cases to cause suicidal depressions). These drawbacks, however, presented themselves to psychopharmacologists as problems to be solved, not as reasons to abandon an exciting step forward in the treatment of mental illness.

Among those inspired by Thorazine's success were a number of institutionally based psychiatrists with the resources to test new drugs as they came to light. Nathan Kline, for example, director of New York's Rockland State Hospital, led pioneering trials with new drugs and was instrumental in introducing a second tranquilizer, reserpine, at almost the same time as Thorazine. Flamboyant and ambitious, he combined groundbreaking research with tireless promotion of both the new drugs and his own role in discovering them. Frank Ayd, a psychiatrist in private practice attached to Taylor Manor Hospital in Ellicott City, Maryland, also organized important early clinical trials and helped advertise the new drugs more broadly by publishing in general-circulation medical journals like the *Journal of the American Medical Association*. Firm believers in drug therapy and willing to buck conventional wisdom to investigate and employ it, Kline, Ayd, and a growing international network of psychiatric researchers shaped the course of the pharmaceutical revolution in psychiatry. Diverse in age, training, and experience, they shared a skepticism of Freudians' claim to have fully explained mental illness and a determination to explore what they felt to be more scientific, and more effective, alternatives.[6]

These psychopharmacology pioneers proved able ambassadors, facilitating the development of their field through conferences and a steady stream of persuasive publications. Perhaps most important, they persuaded the U.S. government to fund their efforts. At the urging of journalist and mental health crusader Mike Gorman, Kline, Ayd, and others testified before Congress about the need for further psychopharmaceutical research. When the Mental Health Study Act was passed in 1955, it provided substantial funds for mental health research and $2 million reserved annually for drug studies through the National Institutes of Health—a monumental sum at the time. The funding was used to create a Psychopharmacology Research Center in the NIH to channel the money and coordinate further research. As historian David Healy puts it, "Biological psychiatry had been capitalized."[7]

It was in this atmosphere of ferment that a second major class of drugs, the antidepressants, came to light. One type of antidepressant was discovered in the 1950s when physicians noticed the improved moods of patients treated with the anti-tuberculosis drug iproniazid. Iproniazid slowly worked its way into psychiatric practice, aided in America by the involvement of Nathan Kline, whose knack for publicity helped bring one of his early clinical trials to the attention of the *New York Times*. Meanwhile, a chemically different kind of antidepressant was discovered by German psychiatrist Roland Kuhn, who had been directed by Geigy Pharmaceuticals to begin a systematic search for new mind drugs. The quest resulted in the discovery that an old compound, imipramine, produced an elevation of mood in depressed patients. Imipramine was announced to American physicians in 1958, when the *American Journal of Psychiatry* published Kuhn's findings.[8]

The discovery of the antipsychotics and the antidepressants represented an extraordinary burst of creativity. Like antibiotics and other postwar wonder drugs, these new medicines empowered physicians to combat many illnesses for the first time, dramatically changing medical practices and contributing to medicine's "golden age." Antidepressants may have come into use only slowly (see chap. 5), but antipsychotics made an impact immediately, all thanks to the intrepid researchers and front-line psychiatrists determined to pursue new hopes even in the face of received medical wisdom and conservative drug companies—which makes it only more telling that both of these classes of drugs were immediately eclipsed in medical and popular worlds by a third class of decidedly less revolutionary, but much more profitable, medicines: the so-called minor tranquilizers.

The first of these minor tranquilizers, Miltown, was discovered in the 1940s by Czech clinical researcher Frank Berger, who had been hired by British Drug House, Ltd., to pursue research in antibiotics. When testing one compound for toxicity in mice, he noticed something peculiar: injected mice lost their fighting reflexes and their muscles relaxed, but they remained conscious and responsive to stimuli. Berger was interested in the drug, mephenesin, but its effects were weak and short-lived. It proved of some medical value, however, and Berger found his way to America after the war and continued to work with the compound. Soon enough a commercial pharmaceutical laboratory on the lookout for new products hired him to become their director of medical research—and to keep tinkering with his discovery. Six years later Berger rewarded their confidence by producing meprobamate, a much more powerful chemical cousin of mephenesin. It remained relatively mild, however, since at least some of its calming power came from its muscle-relaxant qualities rather than from direct sedation of the brain.[9]

The company that hired Berger, however, was not really a pharmaceutical house at all; it was the venerable patent-medicine and toiletries firm Carter Products. And that is where the story gets interesting. Carter Products, founded in 1837, had been famous in the nineteenth century for its fabulously well-advertised flagship elixir, Carter's Little Liver Pills ("Wake Up Your Liver Bile!")—a quintessential example of "snake oil" later regulated out of existence by the Federal Trade Commission. In the 1930s and 1940s, Carter head Henry Hoyt Sr. diversified the family-owned company with a series of new toiletry items including antiperspirants, deodorants, depilatories, and shaving creams. These products were highly successful, owing in part to the company's continued commitment to aggressive advertising: in the 1930s, the public was swamped annually with more than twenty-five thousand Carter commercials over the radio and in the press, the result of an investment of 40 percent of income into promotional activities.[10]

Little in this history suggested a future in prescription medicines. But

Hoyt's search for new products took a new turn during World War II, when he established Wallace Laboratories as an outpost of research and development into respectable therapeutics. This might seem a surprising move for a onetime snake-oil hustler, but the pharmaceutical industry was undergoing enormous changes during the war—changes that would give rise to brand-name prescription drugs and reward upstart companies like Carter with untold riches.

Before 1938, all medicines except for narcotics could be sold legally with or without a physician's prescription. Even before this, however, there were a small number of "ethical" firms that specialized in medicines for physicians and that styled themselves more as part of the medical world than the commercial one. These companies built respectability by advertising only to doctors, thus distinguishing themselves from the proprietary houses that hawked patent (secret ingredient) medicines to the general public (i.e., companies like Carter Products). Because ethical drugs were traditionally sold in bulk form to pharmacists, the pharmacist's skill in mixing pills and elixirs could eclipse the importance of brand names in the pharmaceutical market. And sales of over-the-counter patent medicines dwarfed those of ethical firms. Accordingly, competition in this gentleman's club of firms was restrained, and even physician-oriented advertising was comparatively minimal.[11]

World War II and the discovery of penicillin (first used medicinally in 1940) changed all that. The military needed huge quantities of penicillin, and to ensure this supply, the federal War Production Board promoted vertical integration on a massive scale. Drug companies now began to produce finished products ready for prescription, rather than selling bulk chemicals to middlemen. This raised the value of brand names, just at the time when new kinds of drugs and large-scale production suddenly created the potential for enormous profits. Meanwhile, federal drug regulation had required a physician's prescription for penicillin and other new medicines since 1938. By war's end, there had been a stampede into the prescription drug business. Some chemical suppliers, having become de facto pharmaceutical houses by virtue of delivering finished products to the military, completed the process by buying established ethical houses to handle their distribution and marketing. A number of proprietary firms leaped into the ethical field by expanding or establishing ethical divisions with separate names; and, not least, a number of larger conglomerates bought their way in through acquisitions.[12]

The companies fighting to enter the prescription drug business were not wrong: profits increased dramatically after the war ended. In 1939, total domestic prescription sales were nearly $150 million. Two years after the end of World War II, they had climbed to over $500 million. By 1963, they had surpassed $2 billion. According to government estimates, pharmaceuticals had become one of the most profitable industries in America. The average rate of return for all manufacturing hovered around 10 percent, but by the late 1950s the average rate for pharmaceutical houses was at 20 percent. For drug companies selling tranquilizers, the return rate would range even higher, from 35 percent to 55 percent. Indeed, the tranquilizer market alone was estimated to be in the $100–150 million range by 1956, with manufacturers speculating that the numbers might double in the next year.[13]

When questioned about their high profits, industry spokesmen invariably invoked pharmaceutical companies' standing as medical enterprises, devoted to the discovery of new life-saving drugs. Profits, the argument went, were necessary to fund research and development and to cover the loss of failed drugs. They were also crucial for something else: informing physicians about the uses of new medicines, otherwise known as advertising. Intense marketing would ensure that every physician knew about the brand-name wonder drugs produced by the postwar pharmaceutical system.

As profits continued to rise, advertising and brand competition intensified. Many observers identify American Cyanamid's unprecedented 1948 advertising blitz for its new antibiotic Aureomycin as having inaugurated the new marketing campaigns that quickly became standard practice.[14] Such aggressive sales tactics were far from unusual in the burgeoning consumer culture of the 1950s, when virtually everything had a "pitch." Drugs joined a host of other middle-class goods—from suburban houses and their appliances to new cars and college educations —that were subsidized by the government but were provided commercially through the private sector. In what Lizabeth Cohen has called the postwar "consumers' republic," consumerism was a central institution: it organized and powered the economy, it marked America's political freedom (especially vis-à-vis the Soviets), and it created new opportunities for people to conceive of and pursue happiness and a high standard of living. Advertisers, or the "hidden persuaders," as best-selling author Vance Packard named them, were crucially important to this consumer system, connecting people to new goods and explaining their importance in achieving happiness. They marketed traditional products like soap and automobiles and also expanded into new territories like presidential candidates—and prescription-only medicines.[15]

In fact, prescription drug makers jumped so enthusiastically into the postwar commercial boom that they drew the attention of more than just their potential customers. A series of price-fixing and collusion scandals hit the industry in the 1950s, helping provoke what would turn out to be a half-century worth of nearly constant congressional investigations. These began as early as 1958, when the House of Representatives convened hearings into the advertising of Miltown and its competitors. More serious hearings began in the Senate the following year, when Senator Estes Kefauver, already famed for his televised investigation of organized crime, turned his attention to an industry that, he claimed, earned outrageous profits by hawking medicines like soap. Prescription drug makers still enjoyed a reputation as lifesavers more than commercial giants, but these initial imbroglios indicated both the intensity and the perils of their wholesale leap into brand-name competition and marketing.[16]

Carter Products, then, was not alone among companies reaching for new respectability—and new profits—in the drug business during and after World War II. But company CEO Hoyt had an unusually sharp commercial eye, and one of his hunches played out brilliantly. That hunch, of course, was divining the great commercial potential in Frank Berger's drug mephenesin. Sedatives had always been good business, so this gamble made sense even before the advent of Thorazine and wonder drugs for the mind. When Berger finally produced meprobamate, Hoyt—unused to medical naming conventions—cavalierly trade-named the new drug Miltown after the New Jersey city where Wallace Laboratories was located. His advertisers also helped link the drug to the revolutionary Thorazine by calling it a tranquilizer rather than a sedative, while also distinguishing it as a minor rather than a major tranquilizer with relatively mild action and apparently negligible side effects.

Hoyt must have felt perfectly comfortable in the newly cutthroat atmosphere of brand competition in the pharmaceutical industry. Carter Products, after all, was one of the nation's most storied advertisers. The company happily launched Miltown with its traditional all-out advertising campaign, starting with a special conference for physicians at the Waldorf Astoria hotel and continuing through such tactics as taking out the front cover ad in the *American Journal of Psychiatry* every single month for a full decade.[17]

But Carter was a relatively small-time newcomer to the prescription drug field, and the obstacles to launching a new drug were considerable. The company was not even capable of manufacturing Miltown and had to subcontract out to other drug houses to get the bulk product. It had no connections to physicians, nor did it have the money necessary to hire salesmen (known as "detail men") to create them. In fact, according to Berger, Hoyt briefly got cold feet after initial market testing, and Berger had to drum up interest in the drug himself, which he did by showing a movie of monkeys tranquilized by the drug to other researchers at a conference. Hoyt changed his mind only when American Home Products, one of America's larger corporate conglomerates, offered to pay handsomely for the right to sell meprobamate under their own trade name, Equanil.[18]

Like Carter, American Home Products had initially been a proprietary house, entering the prescription drug business in 1931 by purchasing John Wyeth and Brother's drug house. The new subsidiary was soon strengthened by a steady stream of new laboratory purchases adding to the Wyeth aegis. One of the purchases brought with it patents on a set of innovative production techniques that put Wyeth in a leading role in the production of penicillin during World War II. After the war, American Home continued to expand Wyeth through purchases, bringing in a line of products including antacids, infant formulas, vaccines, and, eventually, tranquilizers.[19] The deal with Carter fit in with this strategy perfectly and promised to benefit both houses: Wyeth won the right to sell a potentially profitable new drug, and Carter got to piggyback on Wyeth's large force of detail men, which could ensure that physicians were aware of the product.

The arrangement proved an unqualified success. While 1954 and 1955 were dominated by Smith Kline and French's Thorazine—by one estimate, it accounted for 100 percent and 99.6 percent of all tranquilizer sales in those two years respectively—the story changed with amazing rapidity the next year. In 1956, Miltown and Equanil leaped from less than 1 percent to nearly 70 percent of new tranquilizer prescriptions (see appendix B).[20] Carter saw its sales nearly triple in the single year after Miltown was introduced, from $15 million to over $40 million. Close to 60 percent of these sales came from meprobamate alone.[21] Meanwhile, American Home saw its profits after taxes nearly double from $90 million to $160 million between 1955 and 1957, eventually reaching $250 million in 1959. Unlike Carter, the giant American Home did not rely exclusively on tranquilizers for sales increases; nonetheless Equanil alone did provide an estimated 10 percent of the conglomerate's sales, an impressive figure considering the total volume.[22]

With such profits in the offing, other pharmaceutical companies eagerly came out with their own minor tranquilizers. By the early 1960s, dozens of mild relaxants were on the market competing with Miltown and Equanil, including the soon-to-be industry leaders, Roche Pharmaceuticals' Librium and its cousin drug Valium. The tranquility bazaar was also crowded by would-be "happiness pills" for the ordinary practice of medicine: older prescription sedatives like the barbiturates, over-the-counter sleep aids, but also powerful antipsychotics like Thorazine, reserpine, and others.

To sell this dizzying new psychotropic cornucopia, drug marketers flooded physicians' offices with armies of detail men, avalanches of direct mail appeals, and a deluge of free gifts, free samples, and free trade journals. One company provided pillows embroidered with the name and catch-phrase of their mild relaxant. Another enticed physicians with a catalogue of consumer goods purchasable through "bonus points" collected by frequent prescribers.[23] Meanwhile, medical journals were awash with tens or even hundreds of pages of slick advertisements pushing trade-named drugs with shameless enthusiasm and salescraft. Most journals were in fact distributed free by virtue of advertising income. Bowing before this reality in the mid-1950s, even the flagship *Journal of the American Medical Association,* under the advice of a marketing consultant, increased its advertising while weakening controls over advertising content.[24] Another brilliant strategy was the *Physicians' Desk Reference,* which first appeared in the late 1950s. The *PDR* was a standard reference volume on pharmacology that helped physicians keep up on new drugs. However useful, it was in many ways a collection of advertisements: drug companies paid by the line to place entries they had written about their own medicines.

Advertisements also featured other classic elements of the Madison Avenue strategic arsenal. Miltown's ads featured a model of "the Miltown molecule" like a sun radiating goodness on the text below, for example, just as Equanil's promised calm seas with its imagery. The campaign for competitor drug Atarax was "previewed" with advance teaser ads. Working off reports in the popular media of shortages of Miltown and Equanil, Atarax's advertisers even went so far as to promise that "every drugstore in the U.S. has now received an initial supply of tablets, and steps have been taken to make sure that supplies do not run short." As early as 1957 both Miltown and Equanil were advertising new dosing forms: Meprotabs introduced meprobamate in a specially coated tablet.[25] Equanil's manufacturers came out with Wyseals the same year: "Now you have a choice of three Equanil tablets," ran the new tagline.[26] There may indeed have been serious medical reasons for the different dosage forms, but in the context of the advertisements, it looked like nothing so much as stylistic product differentiation in a competitive market.

Advertising campaigns were punctuated by sponsored medical conferences designed to meet physicians' continuing-education requirements. In 1964, for example, Carter Products sponsored a conference in New York City celebrating ten years of minor tranquilizers, and the entire proceedings were reprinted in the *Journal of Neuropsychiatry.* In 1978, Roche Pharmaceuticals organized and funded a similar conference at Cornell University called "The Consequences of Stress." These and other conferences featured legitimate medical authorities presenting important findings and usage guidelines, but they also showcased the sponsor drug with far greater frequency than competitor brands, and in unremittingly positive terms. Such affairs could also generate a (carefully orchestrated) buzz in the lay media as well.[27]

Pharmaceutical marketing campaigns, clearly, were no amateur affairs. To sell Miltown, Carter Products turned to Ted Bates and Company, the storied firm that had long advertised Carter's toiletries and patent medicines. Headed by influential advertising guru Rosser Reeves, Bates made a name for itself in the 1950s with its campaigns for Won-

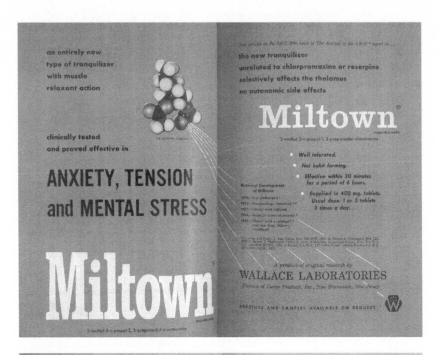

Wondrous new technology promises calm seas for "average patients" in early tranquilizer advertisements. *Source:* Miltown ad, *JAMA,* October 15, 1955, adv. 38–39; Equanil ad, *JAMA,* August 18, 1956, adv. 34A–B. Courtesy of the New York Academy of Medicine Library.

der Bread, M&Ms, Colgate, Anacin, and presidential candidate Dwight Eisenhower.[28] The agency applied its full panoply of tactics to the Miltown case, placing ads in medical journals but also feeding Miltown-related stories to Hollywood gossip columnists and dreaming up a variety of other ways to attract publicity.[29]

Larger pharmaceutical companies had their own in-house advertising departments, and some, like Swiss drugmakers Geigy and CIBA, relied on these departments to produce their marketing campaigns. Most, however, still contracted out to professional agencies, often ones that specialized in medical products. Ads for Roche's Librium and Valium were devised by William Douglas McAdams, Inc., one of the earliest and most influential medical-only agencies. McAdams, founded in Chicago in 1926, began as a traditional agency with accounts for Van Camps Beans, Mother's Oats, and the like. Amid these products it also marketed cod liver oil for E. R. Squibb, the pharmaceutical house. The Squibb account continued to grow until, in 1939, McAdams elected to concentrate exclusively on medical advertising to physicians.[30] Just because the agency focused on professional advertising, however, did not mean it had abandoned its devotion to selling. Its initial campaign for Librium, for example, was hailed in *Advertising Age* for its ubiquity and brilliance, and the drug famously wound up on the pages of *Life* magazine, working its magic on a wild lynx in three full-page illustrations. "New Way to Calm a Cat," the headline announced.[31]

Their powerful commercial backers helped Miltown and Equanil (and eventually Librium and Valium) not only to achieve record-breaking profits but also to capture physicians' loyalties, even in an age of talk therapy. Within a few years physicians were prescribing Miltown and Equanil fifty million times each year—dwarfing not only antipsychotics and antidepressants but all the other wonder drugs of the era aside from antibiotics and the still-popular barbiturates.[32] Their use would only continue to grow over time, reaching a peak in the early 1970s when Valium became the single most prescribed brand-name medicine in America or the world (see appendix B). Miltown and its many competitors might not have been as medically groundbreaking as other new mind drugs, but, for better or worse, for many American physicians and patients they became the face of the psychotropic revolution.

The Minor Tranquilizers in American Medical Practice

Marketing campaigns may have helped power the Miltown phenomenon, but it was physicians who ultimately wrote prescriptions for the new drug. Why were advertisers able to persuade so many supposedly Freudian doctors to embrace a biological therapy for mental and emotional illness? As it turns out, the Freudian medical climate was not so inhospitable after all. Freudians, after all, worked diligently in the postwar years to educate other doctors about anxiety and the importance of treating it early. Drug marketers capitalized on this, portraying their drugs as an invaluable extension of Freudian goals beyond what could be accomplished by the limited number of trained psychoanalysts. And despite being "mind drugs," the tranquilizers were overwhelmingly prescribed by general practitioners rather than psychiatrists. According to regular surveys by the trade journal *American Druggist*, nonpsychiatrists —general practitioners, surgeons, gynecologists, and others—accounted for more than three-quarters of all prescriptions for minor tranquilizers. This proportion was less for the major tranquilizers and antidepressants, but nevertheless substantial.[33]

In some ways, persuading physicians to adopt a new relaxant was bound to be a relatively easy task. Doctors have always relied heavily on sedatives, and a new drug appealed to a profession increasingly aware of the dangers of barbiturates, the main sedative since early in the century. But pharmacy surveys from the 1950s suggest that use of barbiturates did not decline as prescriptions for the minor tranquilizers skyrocketed. Instead, barbiturate prescriptions actually increased temporarily, and then very slowly shrank over the next decade as they fell under increased federal regulation (see chap. 3). According to market observers, the lesson was clear: the minor tranquilizers were not taking over the barbiturate market. They were creating a new one.[34]

This new market did not reflect any newfound belief in biological psychiatry. Ambitious as its adherents were, biological psychiatry was in its infancy. Some starry-eyed early visionaries (who tended, like Frank Berger, to be discoverers of new drugs) did foretell a psychiatry based on brain chemistry, but even supporters like Frank Ayd and Nathan Kline were slow to embrace such radical dreams. Garnering more attention was the work of related specialists focusing not on drugs but on stress itself. Hans Selye, for example, gained a certain respectful audience with his *The Stress of Life* (1956), which argued that stress was a "nonspecific" reaction of biological mechanisms to a wide range of adverse conditions. This wear and tear on the organism, Selye contended, could be observed and measured at the cellular level. Such reasoning accorded with work by psychologists like Raymond Cattell and Ivan Scheier, who measured the physiological manifestations of stress rather than spinning grand theories about its origins.[35]

These still disparate early biological voices, however, had not yet produced a coherent and compelling body of psychiatric theory, and in any case their influence was limited in the medical world as compared with Freud and psychodynamic principles. Drug advertisers, in other words, could not easily look to them for the cultural authority to introduce their products. To understand the Miltown phenomenon one must look elsewhere, in a well developed and highly influential body of medical thought that turned out to be of great use for drug marketers: Freud-influenced psychodynamic psychiatry itself.

Miltown first came to the attention of American physicians with the publication in 1955 of two clinical trials in the *Journal of the American Medical Association* (*JAMA*). In their articles (which ran consecutively), psychiatrists Lowell Selling and Joseph Borrus announced that Miltown had dramatically improved or even cured between 65 and 90 percent of nonpsychotic patients referred to their offices, most of whom suffered from tension, anxiety, and fear states. Both psychiatrists noted Miltown's lack of toxicity or side effects and reported no indications of addiction, concluding that it was "practical, safe, and clinically effective." Medicine's long-running search for the perfect nonsedating, nontoxic relaxant, one suggested, might finally have found its holy grail.[36]

Selling and Borrus presented expansive views of the kinds of medical problems that related to tension and anxiety. Beyond "straightforward" anxiety states, they reported Miltown as effective in treating headaches, mild depression, manic symptoms, "menstrual stress," stomach distress, skin rashes, insomnia, alcoholism, and obsessive-compulsive behavior. In their therapeutic catholicity Selling and Borrus were not alone: indeed, their views were conservative compared with the deluge of reports that followed. Hundreds of clinical trials soon proclaimed Miltown's value in juvenile behavior disorders (bedwetting, "emotional immaturity," delinquency, hyperactivity), cardiovascular disorders, pre- and postoperative tension and anxiety, pregnancy and childbirth, dermatology, allergic conditions, gastrointestinal disturbances, rheumatic conditions, and epilepsy.[37] As one pharmacology textbook observed in 1958, "There is perhaps no other drug introduced in recent years which has had such a broad spectrum of clinical application as has [Miltown]."[38]

The claim that an antianxiety drug could have such diverse and common uses, ironically, owed much to the prominence of Freudian psychiatry. During World War II, thousands of otherwise healthy soldiers—including noncombatants—had succumbed to "shell shock," lending credence to the psychodynamic argument that severe stress and anxiety could produce mental illness in virtually any person. As they returned from the service to assume leadership positions in their profession, military psychiatrists brought with them a war-forged vision of a preventive therapeutics focused on the early treatment of stress-related disorders. As the profession reoriented to "community-based" care of patients with such relatively mild symptoms, more psychiatrists elected to enter private office practice, and asylum psychiatrists dedicated to care of psychotics shrank in numbers and prestige.[39]

This heightened attention to nervous symptomatology made it easier for new populations, particularly the burgeoning white-collar middle classes, to imagine their emotional and mental distress as formal illnesses, treatable by medicine. Postwar psychiatry, in turn, equipped with psychodynamic arguments, was ready to take them seriously.[40]

Anxiety sat at the heart of this new psychiatric paradigm. Central to Freud's theories and also clearly important in wartime instances of shell shock, *anxiety* was a term to conjure with in virtually every version of postwar psychodynamic psychiatry. The American Psychiatric Association's first *Diagnostic and Statistical Manual* (1952), for example, identified anxiety as the "chief characteristic" of the neuroses (a catchall term referring to virtually all nonpsychotic mental illnesses). Echoing Freud, the *Manual* explained that neurotic symptoms such as phobias, depression, obsessive or compulsive behavior, dissociative behavior, and others reflected a person's dysfunctional efforts to cope with anxiety. Treatment involved recognizing and coming to understand the source of the underlying anxiety.[41]

The emphasis on anxiety was particularly strong in the branch of psychodynamic medicine known as psychosomatic or psychophysiologic. These terms referred to physical conditions—skin rashes, stomach aches, heart palpitations—that, according to laboratory tests, had little or no organic basis. Psychosomatic theory held that such conditions were either symbolic conversions of inner turmoil (thus a blinding headache might be a repressed wish to murder the boss), or true organic conditions arising from a constant bombardment of stress. The most common kinds of symptoms discussed as psychosomatic involved skin, stomach, and heart troubles, but during the height of this approach, the symptomatology knew virtually no limits. The *Manual,* for example, included sections on psychosomatic problems in virtually every bodily system from blood to bones to sensory organs. Such claims were not lost on Miltown prescribers, many of whom appear to have prescribed the drug routinely to ease patients' worries about their physical illness. As one psychiatrist explained, *all* organic diseases had "a degree of functional [psychological] overlay, small in some and large in others."[42]

Given the extensive reach of anxiety and its infinitely related ills, it is hardly surprising that when America's mental health professionals peered out into society at large, they saw a seething mass of pathology. Major epidemiological studies found that astonishingly high proportions of noninstitutionalized Americans—ranging from 10 to 25 percent—were "psychosocially impaired" according to standard mental health scale criteria.[43] The *National Disease and Therapeutic Index,* a medical industries market survey, tallied tens of millions of physician visits for neuroses, personality disorders, and psychosomatic conditions, placing such psychiatrically oriented problems among the most common reasons for seeing a doctor.[44] An influential study at Mount Sinai Hospital in New York found that over 80 percent of one thousand so-called problem patients—those presenting "puzzling diagnostic features"—turned out, upon closer examination, to have "psychological factors as the basis for their complaints and illness."[45] By the mid-1960s, conventional wisdom held that up to half of all patients seen in general practice were free of organic illness, their suffering entirely psychological.[46]

Faced with such a deluge, American psychiatry seemed woefully understaffed. Trained analysts remained a tiny population: the American Psychoanalytic Association's membership roster held only 695 names in 1957 and a decade later had barely doubled with 1,300.[47] Meanwhile, the

Joint Commission on Mental Illness and Health, created by Congress in 1955 to assess national mental health policy, estimated that fewer than 5 percent of all physicians were psychiatrists. Their ranks, although growing, were still spread thin: there was only one psychiatrist, on average, for every eighteen thousand Americans, and in some southern areas the ratio sank as low as one for every thirty-four thousand. The Joint Commission's influential final report, issued in 1961, called for a "national manpower recruitment and training program" to augment the overburdened mental health professions.[48]

As most psychiatrists readily admitted, general practitioners served as the front lines in any battle against mental illness, since they saw patients when their troubles were still relatively mild and easily treatable. Indeed, a central goal of postwar American psychiatry was to persuade other physicians to take the mental and emotional dimensions of health seriously, rather than simply turning away patients whose troubles were "all in their head." But how were general practitioners to treat their tense and anxious patients? Psychotherapy was time consuming, and even biologically oriented psychiatrists like Nathan Kline warned that "amateur psychodynamic analysis is just as dangerous as amateur surgical intervention."[49]

To marketers of Miltown and other antianxiety drugs, this medical dilemma represented a golden opportunity. It gave them a potential entrée into a medical world still focused on Freud and talk therapy. Even better, that entrée was to the most attractive part of the potential market: not the relatively few numbers of the seriously ill but the millions of "walking wounded" who were distressed but still largely functional. Targeting this vast but amorphous population could lead to immense sales while avoiding direct competition with the Freudian establishment.

In keeping with this strategy, drug advertisers echoed Freudians in emphasizing the ubiquity of anxiety even among otherwise healthy people—especially in general-circulation journals like *JAMA*. Early Miltown ads, for example, promoted the tranquilizer to treat "mental stress" or "tension" in "the average patient in everyday practice" and deployed psychosomatic reasoning to suggest the drug for allergies, arthritis, asthma, and other problems.[50] In the *American Journal of Psychiatry*, meanwhile, Miltown ads claimed that the drug "facilitates psychotherapeutic rapport."[51] New competitor Roerig Laboratories (run by Pfizer) announced its minor tranquilizer, Atarax, with an ad campaign in *JAMA* suggesting no fewer than twenty-five adult and ten pediatric indications for use, including "tension," "nightmares," "temper tantrums," and "enuresis" (bedwetting).[52] Equanil's marketers, meanwhile, in a series of ads featuring the tagline "anxiety is part of every illness," quoted generously from physicians linking anxiety to hypertension, allergies, gastrointestinal disorders, peptic ulcers, and "hypochondriasis." Each page repeated the mantra: "In every patient . . . In every illness."[53] Other Equanil ads in *JAMA* appealed more directly to Freudian reasoning:

> The need for equanimity is universal because anxiety and tension arise from universal forces—conflict, aggression, dissatisfaction, frustration. As a response to destructive affects, anxiety and tension are the commonplace of the age. They are force and subforce in every medical practice . . . with [Equanil], the physician has a long-sought means to counter anxiety and tension—to promote equanimity.[54]

These campaigns made certain that Freudians' lessons about taking anxiety seriously would percolate through medical circles, even to ordinary general practitioners oblivious to the latest psychiatric research. Friendly and accessible, inviting with gorgeous graphics, and largely free of complex psychiatric terminology, the ads translated psychodynamic psychiatry's obsession with anxiety into simple life problems that physicians confronted every day in their offices. And, most important, they promised a safe and effective way to do something for patients suffering from those problems—a way clearly defined as proper medical therapeutics, supported both by the impressive lists of citations in the advertisements and, after 1962, by an FDA-mandated technical summary of tested indications, efficacy, and side effects.

Many physicians were clearly ready to embrace such appeals. "Certain limitations inherent in psychotherapy have at times hindered its proper application by generalists," explained one general practitioner tactfully, but "tranquilizers are a tool that could be used by the psychiatrist and generalist alike."[55] If generalists could not psychoanalyze their troubled patients, they could at least ease worries with a pill, possibly preventing a minor condition from worsening into serious mental illness. As one Miltown supporter explained, the drug "may help most significantly in stopping, at the anxiety stage, the psychodynamism toward disaster." This "little breathing spell," he asserted, "may be life-saving indeed, not only to individuals, but to social groups as well."[56] Given the acknowledged epidemic of anxiety in America, such triage was of no small importance. "There is no longer an opportunity to choose whether or not we will care for these people," instructed one generalist. "We can't afford to ignore them. There are too many neurotic personalities and too few psychiatrists to treat them."[57] Apparently agreeing with this sentiment, a welter of articles appeared in general medical journals calling on readers to learn about the drugs and how to prescribe them—a call well heeded, to judge by prescription figures.[58]

Miltown's popularity among nonpsychiatrists did not necessarily mean a parting of ways with Freudian ideas. Consider, for example, a 1957 report by allergist Ben Eisenberg in the *Journal of the American Medical Association*. The patient was a 35-year-old female schoolteacher suffering from a "marked sensitivity" to feathers, house dust, wheat, and egg whites. Simply avoiding these allergens had helped somewhat, but her skin rashes occasionally returned, and she still slept poorly. Eisenberg suspected that the remaining problems might be psychological in origin, but "despite a show of tears at almost every visit," the patient refused to accept this possibility. "I've read where allergies can be triggered by the emotions," she said, "but mine can't be! I never worry, have a happy home life, wonderful husband, and most of the things I need." Eisenberg disagreed, finding her "quite obviously . . . depressed, tense, and anxious." He prescribed the antipsychotic Thorazine, and when that produced "no favorable result," he tried reserpine, with the same lack of results. Finally, he gave her Miltown, and under its calming influence she finally revealed her feelings of guilt about having placed a severely handicapped daughter in a state institution. "After several interview sessions, during which she unburdened herself of long-hidden guilt feelings concerning this child," Eisenberg concluded, "the dermatological condition became much improved."[59]

With its neat, clichéd resolution, this case history reveals the difficulty in separating Freudian and biological psychiatry in the 1950s. Eisenberg used drug therapy, but he believed that his patient's illness had been caused by a scarring past experience, not a derangement of her brain chemistry. The drugs he prescribed were not to cure her condition but merely to lessen her anxiety to the point that she could actually face the origins of her problems and work through them. The overall trajectory remained archetypally Freudian: puzzling symptoms turned out to be outward signs of an inner psychic conflict, and cure came about only when the patient fully understood this. But in practice, the doctor had treated her with Miltown.[60]

Marrying psychodynamic to psychotropic practice in this manner appears to have been quite common within both psychiatry and medicine at large. Conventional wisdom, repeated in countless textbooks and instructional articles in medical journals, held that psychiatric drugs did not cure but allowed "only a degree of symptomatic relief."[61] This symptomatic relief, however, could be invaluable, even as a psychiatrist embarked on the more profound task of treating the underlying pathology.[62] Indeed, according to early surveys, the overwhelming majority of American psychiatrists employed drugs in their practices, even if their stated preference was for talk therapy. They even used drugs in office practice for milder conditions such as the neuroses—disorders for which drugs, according to the survey respondents themselves, were of little value. "Even when psychiatrists reported that they 'never or rarely' used drugs," one study concluded, "both general psychiatrists and psychoanalysts listed a much more frequent use." Psychiatrists, even psychoanalysts, used drugs "more than they apparently realize."[63]

What could have been perceived as a weakness—the inability to cure underlying mental illness—thus turned out to be a great boon for Miltown, allowing even skeptics to use the drug adjunctively in psychiatry and as a superior relaxant for those in need of temporary relief. Many physicians apparently agreed that Miltown was helpful for the tensions of otherwise healthy men and women struggling with normal life challenges. In the general-circulation journal *Science,* for example, one reviewer recommended Miltown for "the 'normal population': a businessman with a demanding and unreasonable supervisor or a woman with insufficient funds to run her home according to her ideal standard."[64] In medical journals, it was not uncommon to read about the value of minor tranquilizers for such situations as bereavement, failure to get a job or a promotion, premarital jitters, the "need to match wits with a competitor," fear of public speaking, resentment, divorce, or the tensions of "a lawyer involved in preparing a complicated legal brief or an overworked executive carrying a tremendous burden of responsibility."[65] As advertisers surely knew, such reasoning expanded the range of indications for the minor tranquilizers almost without limit. Between anxiety, the neuroses, the psychosomatic ailments, and the "ordinary" tensions of living, practically every American might have merited tranquilizing at some point during a given year.

All this is not to say that advertising dictated physicians' use of tranquilizers. Although studies have consistently shown that ads do influence prescribing habits, the relation between advertising and medical practice is complex.[66] Medical practice and advertising narratives built on each other, working together in this early era to produce more prescribing for more kinds of mild complaints.

The results of this medical and commercial synergy were extraordinarily beneficial for Miltown, according to two early studies of drug prescribing completed in the late 1950s. Surveying prepaid group health insurance plans in New York and Washington, D.C., both studies found that Miltown was prescribed approximately one-third of the time for psychiatric disorders (primarily "ordinary" tension and anxiety), with the rest dispensed for an array of other medical problems ranging from heart conditions to muscle pain to skin problems. Interestingly, one of the studies revealed that over 10 percent of all Miltown prescriptions were written for no stated reason.[67] This suggests—perhaps unsurprisingly, given its astonishing range of uses—that Miltown had, for some physicians at least, achieved a sort of aspirin-like status, universal and commonly accepted, but also removed from the actual project of curing so that its purpose might go unstated. Within a few short years, Miltown had become a virtual panacea, offered like patent medicines of old for nearly any ailment.

The Successor to the Tranquilizers: The Decline of Miltown and the Rise of Librium and Valium

Miltown's good fortune was not to last for long, however. Within a year of its introduction, physicians had already begun to notice a wide range of side effects, including possible addiction (see chap. 3). The drug proved to have sedative qualities after all, often leaving patients drowsy, and higher doses could be toxic, as attested by a growing number of suicides by individuals taking Miltown.[68] Meanwhile, partisans of the newly emerging gold standard for testing medical outcomes—the double-blind placebo-controlled trial—pointed out that little reliable evidence for Miltown's efficacy had been produced in the first place.[69] The widely lauded (but often ignored) *Medical Letter on Drugs and Therapeutics,* a noncommercial journal featuring skeptical, unsparing reviews of drugs, summarized their judgment in these words: "Apart from a placebo effect . . . there is little reason to prescribe [Miltown] at eight cents per 400mg tablet in preference to phenobarbital [a barbiturate], which is much cheaper."[70]

In the face of such discouraging news, and amid a parade of new, well-advertised competitor drugs, use of Miltown and Equanil fell off dramatically in the early 1960s. In 1964, deemed to be more expensive but not an improvement over barbiturates, they were struck from the U.S. Pharmacopeia (a national listing of drugs of tested value and worth). By 1971 Miltown, though it now accounted for a meager 6 percent of all minor tranquilizer prescriptions, was featured in the *American Journal of Psychiatry* as a case study in irrational prescribing.[71] In the end, Miltown's moment in the sun was so brief that one leading pharmacology textbook, Goodman and Gilman's *The Pharmacological Basis of Therapeutics,* missed it entirely: the 1955 edition was too early to include it, and the next edition in 1965 barely noticed it amid the welter of competitors.[72]

The decline of Miltown, however, did not mean the end of the minor tranquilizer phenomenon. These new drugs had carved an enduring therapeutic niche in American medical practice, one that proved far longer lived than any individual medicine. Indeed, minor tranquilizer use would continue to rise dramatically through the early 1970s, maintaining their overwhelming dominance over other psychiatric medications and their overall position among the nation's most popular medicines. At their peak in the mid-1970s, well over 100 million prescriptions were filled for the minor tranquilizers yearly. By contrast, antidepressants had climbed to only 30 million prescriptions per year and antipsychotics to 25 million, while the older "mind drugs," barbiturates and amphetamines, were in the midst of long-term declines at 50 million and 6 million prescriptions annually (see appendix B).[73]

Minor tranquilizers owed much of their continued popularity to the success of Miltown's famous successor drugs, Librium and Valium. These drugs were discovered in the wake of the Miltown sensation by chemist

Leo Sternbach for the Swiss pharmaceutical house Hoffmann–La Roche and became available to physicians in the early 1960s, just as Miltown's reputation had begun to tarnish.[74] These drugs added even stronger muscle-relaxing power to their tranquilizing effects, which helped them calm patients without sedating them. As one psychiatrist noted approvingly in the *American Journal of Psychiatry,* Valium "calmed patients and improved moods without interfering with mental alertness, work drive, or positive affective responses, and without producing inappropriate elation or undesirable behavior."[75] A report in the *Journal of the American Medical Association* hailed Librium as "certainly more specific" in its "antianxiety effect" than Miltown or the antipsychotics.[76]

Like their predecessors, Librium and Valium quickly gained a reputation for combating a breathtakingly wide range of illnesses. Studies soon sprang up advocating the drugs for every kind of ailment previously treated with Miltown, plus whole new fields like sports medicine opened up by their stronger muscle-relaxing qualities. (In a testament to the excitement, and the early confusion, surrounding the new psychiatric drugs, a number of researchers erroneously identified Valium as an antidepressant, too.)[77] "Since its introduction," commented a psychiatry textbook with careful neutrality, "it has been claimed that Librium has achieved good results in practically every disorder."[78] After describing Valium's many and varied therapeutic qualities in the medical journal *Psychosomatics,* one physician paused to ask (only half rhetorically), "When do we *not* use this drug?"[79] In the late 1960s and early 1970s, when Valium and Librium reigned as the first and third most prescribed drugs in America, respectively, one had to wonder.

Such cure-all claims would have sounded familiar to Miltown prescribers, and for good reason. They borrowed from the same Freudian reasoning about anxiety that Miltown had and were circulated through the same intense marketing system. Indeed, Librium's advertisers made such a big and successful splash that *Advertising Age* ran a feature analyzing it as a textbook case of new product differentiation. For the most part, the campaign echoed standard promises of relief for the widest possible range of "everyday" patients. Librium's advertisers highlighted their appeals with a Madison Avenue strategy called the "unique selling proposition," emphasizing a single distinctive feature of their product: it was a genuinely new chemical, unrelated to existing tranquilizers. This did not necessarily mean it was better than existing tranquilizers, but initial marketing campaigns nonetheless portrayed Librium as a revolutionary advance over earlier drugs ("the successor to the tranquilizers") and emphasized its chemical uniqueness ("not a manipulated molecule"). "How does a pomegranate taste?" asked one ad.[80]

Despite their immense success, Librium and Valium were not without their detractors. The *Medical Letter*, for example, initially disapproved of them only marginally less than it had Miltown, and some observers leveled familiar criticisms against flawed or unscientific research trials.[81] Compared with Miltown and other competitors, however, they were surprisingly nontoxic, relatively free of side effects, and highly effective. Physicians had come to rely on minor tranquilizers, and Librium and Valium seemed to be the best available. In the late 1970s, Americans in a far different medical and cultural climate would look with greater suspicion on these drugs, but in the intervening years Librium and Valium remained the most resounding of psychopharmacology's triumphs.

Pursuing Happiness: The Citizen as Patient

For all their power, pharmaceutical companies and physicians could not create the Miltown phenomenon alone. For that, they needed the cooperation of innumerable patients and potential patients. And just as physicians needed to be "educated" about anxiety and the new drugs, so too did their patients. Some of this education was standard outreach of the public health variety, like the establishment of community mental health centers in urban areas. But the postwar era also introduced a wholly new element: prescription drug advertising and public-relations campaigns designed to escape professional medical circles and circulate in popular media. These campaigns invited Americans to think of themselves as consumers in a new commercial bazaar for the psyche. Advertisers had long sought to sell self-fulfillment through proxy products—to sell happiness through a new car or beauty through soap. Now they had an opportunity to cut out the middleman and literally sell "happy pills."

The "happy pill" campaigns ran the gamut of Madison Avenue techniques. When Senator Kefauver investigated the drug industry in 1960, his researchers found that pharmaceutical companies maintained public-relations machinery that extended their advertising message well beyond medical journals. They provided "educational" interviews with company spokesmen about the virtues of the latest wonder drug, for example, and then offered them as free public-service announcements for radio stations.[82] They took care to ensure that company experts were always available to answer reporters' questions and cast the drugs in a flattering light. And if a celebrity used Miltown or the "Miltini" (martini and a Miltown) became "all the rage" in Hollywood, they made certain that gossip columnists wrote about the new fad.[83] Carter Products' quest for headlines included commissioning a new sculpture from famed surrealist Salvador Dali. *Business Week* dutifully ran a picture of the piece, "Release from Anxiety," in its report on the conference where Carter unveiled it.[84] Roche Pharmaceuticals even ran advertisements for Librium in special copies of *Time* magazine mailed to doctors for use in their waiting rooms; the doctors were supposed to tear out the ads before putting them out for patients to read.[85]

Even when critics like Senator Kefauver raised concerns about drug advertising, they served as just one more avenue for public dissemination of marketing information. Kefauver's hearings, for example, and the long succession of congressional investigations that followed, were closely tracked in the news media. The resulting exposés often featured quotations or even full recreations of ads and promotional rhetoric, circulating them to whole new audiences. In 1957, for example, *Woman's Day,* *Reader's Digest,* and *Fortune* magazine all criticized a leaflet packaged with Pfizer's new minor tranquilizer Atarax, which described the drug as useful for, among other situations, interviews, competitive exams, public appearances, "tension and excitement in noisy environments," and even the anxiety stemming from "differing opinions." Although they cited the leaflet disapprovingly—"This list might be funny if it were not frightening," one remonstrated—they also informed readers about Pfizer's marketing claims for the drug.[86]

Advertisers could also get headlines by dint of creative or catchy marketing campaigns. Librium's advertisers, for example, had great success with an early-1960s campaign about tranquilizing wild animals: *Life* magazine devoted three pages to illustrating the effects of Librium on a wild lynx, and fellow Luce publication *Time* also covered the "new way to calm a cat" story. In fact *Time* went even further, faithfully reporting Roche's claim that Librium "comes close to producing pure relief from strain without drowsiness or dulling of mental processes."[87] Few reports on new drugs so closely mimicked advertising campaigns, but many did "educate" readers by listing new drugs by brand name and manufacturer and briefly summarizing the salient marketing claims.[88] It is certainly no surprise to find mass-market journalists excitedly reporting new medical and technological advances, especially in an era famed for revolutionary wonder drugs. But it is important nonetheless to recognize the role drug advertisers had in instigating and shaping this coverage.

Some of the marketing ploys went well beyond subtle orchestration. According to Senator Kefauver's researchers, some reporting in popular media had actually been written by employees of the Medical and Pharmaceutical Information Bureau, an industry public-relations outfit. Such articles could be even more extravagant than the advertisements themselves. A feature in *Cosmopolitan*, for example, praised Miltown for relieving stomach distress, skin problems, "the blues" and depression, oversensitivity to summer heat, fatigue, inability to concentrate, lack of "social ease," behavior problems in children, and insomnia. All this curative power, moreover, came "free of penalty" because Miltown "is not habit-forming, and tolerance does not develop." In a follow-up article, the same magazine explained that the drug had earned the name tranquilizer because it produced "perfect peace or calmness of mind." A new eye-popping list of uses ensued, calibrated to address popular issues of the day such as sexual "frigidity" and juvenile delinquency: Miltown could "banish 'that tired feeling,'" make workers "happier and more productive in their jobs," help "frigid women who abhorred marital relations [to] respond more readily to their husbands' advances," and transform youngsters who "seemed headed hell-for-leather into juvenile delinquency" into "calm . . . quiet, cooperative, better-behaved children, at last able to talk about their problems and eager for help in overcoming them."[89]

Well beyond the pages of medical journals, then, consumers of popular media were receiving an intensive education from drug advertisers that complemented Freudians' campaign to raise awareness of anxiety. The messages also harmonized with the broader consumerist ethic of the postwar period, in which Americans were taught that expanding consumer choice was a personal, political, and economic virtue. And yet potential patients were not simply pawns of drug advertisers, tricked into imagining anxiety and emotional suffering that did not actually exist. Even the white middle classes comfortably ensconced in prosperous suburbs, colleges, malls, and other key locales of what historian Lizabeth Cohen calls the "Consumers' Republic" were not guaranteed freedom from emotional hardship. Not everyone lived happily in the officially encouraged "nuclear" suburban family; parents did genuinely worry about juvenile delinquency and teenage pregnancy; and white-collar work could produce anxiety and stultifying conformity as well as good wages.[90] Medical authorities and drug advertisers circulated particular stories about what these and other miseries meant, but they did not invent the real and often painful experiences that prodded people to pay attention to those stories. There can be little doubt that evangelists for both Freud and Miltown shaped many Americans' understandings of their own problems, but ordinary people actively used the information they received to pursue their own ideas of self-fulfillment.

Indeed, by all available evidence many people eagerly sought out the new drugs, just as they pursued other techniques for coping with the cosmic and everyday stresses of their lives. (Norman Vincent Peale, author of perennial self-help bestseller *The Power of Positive Thinking* [1952], knew this well.) Obviously they would not have done this unless they had first heard about the new "happy pills," perhaps even from an industry-inspired source. But they nonetheless brought their own agenda to their meetings with physicians, asking for or avoiding certain medicines, and making personal decisions about how (and if) they would take the medicines once they had been prescribed. Indeed, the majority of prescriptions filled for the minor tranquilizers required no meeting with a physician at all: most prescriptions, once obtained, were indefinitely refillable in these early years. This gave patients a great deal of choice in how they used their drugs.

One important piece of evidence testifying to patients' willingness to use their power of choice was the ubiquity of physicians' commenting (or, more commonly, complaining) about it. Many physicians reported patients upping their doses or, upon feeling better, cutting them down or even stopping entirely, all without consulting their doctor.[91] Ben Eisenberg, the allergist mentioned earlier who used Miltown to treat the schoolteacher's skin rashes, noted that patients helped by the drug continued to take it on their own as needed, and that "several . . . hoarded part of the tablets against future need, just in case their pharmacy 'ran out.'"[92] Prescription surveys also consistently ranked the minor tranquilizers as the drugs most likely to have been purchased with a refill prescription: refills accounted for over half of all sales.[93]

Beyond deciding how many pills to take and how often to take them, some patients also used Miltown's simplicity and effectiveness to exert control over larger therapeutic issues. One salesman, for example, described in a 1955 article on Miltown in *Northwest Medicine,* called his psychiatrist to cancel all future therapy sessions after several days on the drug, "saying that, for the first time in his life, he was free from a constant feeling of shaking and tension and that he was going back to work."[94] In another case, reported in *American Practitioner and Digest of Treatment* in 1958, a woman used Miltown to resist pressure to quit her job for health reasons. Although the divorced mother had had headaches since the age of 11, her physicians traced her current problems to her recent promotion to an executive position at work. Miltown eased her headaches, leaving the woman "happy in her work with less tension than in many years." But, the physicians added, "thus far the patient has failed to reorganize her life to eliminate the basic conflicts underlying her tension, despite repeated warnings that meprobamate therapy does not cure, but only aids." Despite her physicians' clear belief that she ought to "reorganize her life"—presumably to end her dual role as business executive and mother—this patient made the decision to use Miltown to handle her challenging responsibilities.[95]

Not all physicians were happy about their patients' efforts to control their own therapy. Noting the "tendency for relatives and friends to trade these products back and forth," one physician warned readers of *Postgraduate Medicine* that "the layman would hesitate to diagnose acute appendicitis in himself, but has no compunction about diagnosing an anxiety state in himself or in a friend and proceeding with the treatment."[96] Many physicians blamed the popular media for such cavalier attitudes, and for the more irritating problem of patients demanding minor tranquilizers or other drugs from their doctors. As one put it, "The patient with the *Reader's Digest* in his hand is one the doctor does not always handle skillfully."[97] Immersed in a press "teem[ing] with articles giving the impression that the elixir of happiness has been found," one physician wrote, an "eager and gullible public . . . puts its physicians under pressure to supply happiness."[98]

Evidence of physicians' uneasiness with patient demand and other control issues can also be found in Carter Product's marketing strategies. Miltown was by far more famous than Equanil, but within a year or two Equanil had taken a commanding lead in the competition for prescriptions.[99] According to market observers, Miltown's very fame was the problem: physicians, jealous of their craft, did not like patients knowing what was being prescribed to them, and certainly did not want to compete with *Cosmopolitan* for prescribing wisdom. The case of "mind drugs" was especially delicate, since not every patient would be happy to learn that their supposedly physical illness might actually reflect mental instability. Carter Products ultimately introduced a new version of the drug, the sustained-release capsule Meprotabs, which they advertised primarily as a way to prescribe Miltown without the patient knowing it.[100]

If by changing brand names Carter hoped to limit popular challenges to medical authority, however, it was a vain effort. And this is one of the central ironies of the vast new medical-industrial complex. Commercialized medicine arrayed enormously powerful forces to encourage the use of new blockbuster drugs like Miltown, but it also expanded public discussion of medical practices in new and ultimately uncontrollable ways. Expanded "health" sections in popular magazines and other forums, for example, were crucial sites for spreading information about the latest drugs, but they also invited public debate in ways that private physician visits did not. Advertisers' public bullhorn gave them a chance to frame these popular debates, but it did not give them control over patients any more than it had given them control over physicians.

Moreover, the new cultural spaces opened up by commercialized medicine were designed to attract the interest of new populations who did not necessarily think of themselves as having problems with anxiety. This too could be a double-edged sword. It cracked open potential new markets, but it also displayed the spectacle of the commercialized psyche in popular media for all to see—not just those in search of medical relief. And if the Miltown phenomenon generated a great deal of enthusiasm, the prospect of eliminating anxiety with a well-advertised and popular "happy pill" could also be profoundly disturbing. That America seemed to be drifting toward what one critic called "total tranquilization" was, for some observers, itself something to be quite nervous about.

CHAPTER TWO

Listening to Miltown

"HAPPY PILLS." "Peace pills." "Peace-of-mind pills." "Don't-give-a-damn-pills." Whatever the catchy name, the minor tranquilizers were everywhere in the popular media in the years after their discovery. Astounding new wonder drugs seemed to promise unprecedented control over mind and emotion, heralding radical changes in the nation's psychic landscape. Miltown's name may no longer be remembered, but during its moment in the sun it served as the first popular face of this psychotropic revolution. Would the tranquilizer and its cousins truly bring about a "choose your mood society," as *Fortune* magazine reported? Had consumer society matured to the point that the best guide to true happiness could be found not in the Bible but in the *Consumer Reports* review of new drugs? Did the existence of a "happy pill" require a rethinking of human nature and identity? Through hundreds of articles in magazines from *Catholic Home Journal* to the *New Yorker*, in newspapers, and on television, Americans in the 1950s and 1960s confronted the implications of Miltown long before Peter Kramer advised them to "listen to Prozac."

If "happy pills" and "choose your mood society" sound more like advertising slogans than sober medical pronouncements, there are good reasons for it. With enormous resources to devote to educating both physicians and the public, drug marketers had unique power to shape the cultural landscape. They translated medical language into the simple stories of American consumer culture, transforming "wonder drugs" into "blockbuster drugs" and spinning utopian tales of a benign commercial conquest of the psyche. Important as they were, however, they were not the only voices shaping the psychotropic phenomenon. Their success in projecting Miltown and its cousins into the public eye invited others to participate too, people who saw opportunities to "use" the ubiquitous new drugs to pursue their own cultural and political agendas. The result was a dynamic public dialogue about the meaning of wonder drugs for the mind, deeply conditioned but never fully dictated by the forces of commercial medicine.

This chapter examines the first public controversy over the new mind drugs, which followed fast on the heels of Miltown's introduction. In later decades, such controversies would focus on women's use of what had become the quintessential "women's drugs." But in the late 1950s, tranquilizers had not yet become women's drugs; rather, with advertisers' help, they shared a popular spotlight with antibiotics and other revolutionary wonder drugs. Indeed, Miltown first attracted widespread public criticism because of the supposed threat it posed to the virility of white-collar men. Amid broad cultural campaigns to reinstate traditional gender roles in postwar America, white-collar men were already suffering what many critics called a crisis of masculinity. They were, so the story went, becoming "soft" in the comfortable world of consumer affluence and the group-think of middle management. The much-hyped and highly visible Miltown became a perfect symbol of these threats. In raising a moral panic over its widespread use, gender traditionalists championed what they depicted as older, more authentic kinds of masculinity centered around an animal vitality powering men's ambition, competitiveness, and rightful social authority.

Advertisers were not willing to give up on male customers, however, responding with new campaigns promising that drugs would restore, not damage, vital masculinity. In later decades images of women in drug advertisements would provoke intense feminist criticism. But fully half of all drug advertisements featured male models or otherwise targeted male patients, and these advertisements received virtually no criticism or analysis. Critics, even feminists, assumed that they were the normal standard by which ads targeting women ought to be judged. Ironically, however, it was through these representations of men that advertisers accomplished some of their most important cultural work in the 1960s. It was these images that helped circulate influential new biological narratives of masculinity. The advertisements became a meeting ground for medical and scientific literature, on the one hand, and mass-media cultural criticism, on the other—a place where hybrid concepts of selfhood incubated and then reverberated widely in both popular and medical cultures.

The chapter concludes with a brief preview of how and why the tranquilizers eventually became "women's drugs": not just because women used them more, but because feminists like Betty Friedan adopted them in the 1960s as a political issue just as masculinist critics had in the 1950s. Their efforts would come to fruition only in later decades, but even at that early date the implications were clear. When Americans "listened to Miltown" and its cousins, they heard not just the pills but the people who packaged them—including such decidedly nonscientific actors as drug advertisers and middle-class cultural critics. Commercial medicine framed the debates, gathering new cultural power for marketers and the medical-industrial complex, but also inviting other voices to contest medical wisdom about the psyche and identity. This chapter traces this process, shedding light on the cultural interactions, conflicts, and unlikely collaborations through which physicians, advertisers, activists, journalists, patients, and others created and advertised the message in the pill bottle.

The Age of Affluence Meets the Age of Anxiety

Before delving into the debates surrounding Miltown, we need to take a detour into the cultural history of a closely related subject: anxiety, the illness Miltown and the tranquilizers were supposed to vanquish. Anxiety belonged to a rich tradition of "nervous illnesses" that predated the era of Freud, stretching back at least to the late nineteenth century. These maladies were widely thought of as illnesses of affluence, problems to which elites were especially susceptible because of their refined and complex psyches. They offered medical proof of elites' distinctiveness as a class, while at the same time signaling a class-wide psychic crisis, usually attributed to a breakdown in "natural" gender roles. This made them very useful for cultural campaigners interested in reforming white-collar life: the reforms became a medical as well as a political necessity. The 1950s "epidemic" of anxiety and neurosis belonged in important ways to this ongoing cultural dynamic. To understand popular responses to Miltown, therefore, we must first spend a few pages with the nervous illness tradition.

We begin in the late nineteenth century with neurasthenia, a briefly fashionable and surprisingly influential ailment that is perhaps the best-documented chapter of the nervous illness tradition. According to eminent neurologist George Beard, those most at risk for neurasthenia were Anglo-Saxon people of refined sensibilities and high intelligence, often "brainworkers." Far removed from the animal vigor of their ancestors and of less evolved races, these elites did not have the nervous resources to cope with the pace and intensity of industrializing America. As they "depleted their nervous energy," Beard wrote, they were struck by bizarre and mysterious symptoms ranging from exhaustion to "inebriety" (drug and alcohol abuse). Anyone might suffer from one or more of these symptoms, of course, but the diagnosis of neurasthenia was largely reserved for those who could afford to see the neurologist.[1]

Diseases have always lived double lives, with dread epidemics typically being used to stigmatize the poor, immigrant, and nonwhite groups who disproportionately suffer them.[2] Neurasthenia represented that dynamic in reverse: it seemed to provide medical evidence of the mental superiority of the native-born and well-to-do. According to the medical and popular logic that surrounded the disease, only civilized men and women of high intelligence and complex inner lives were delicate enough to suffer from it. On coarser and less refined souls, it would barely register.

Distinguished or not, however, neurasthenia remained an illness—an indication that something was amiss with America's white native-born elites. The nation's best men had become too civilized, critics like President Theodore Roosevelt charged, and needed to be revitalized through active, daring contact with nature, hard work, sports, hunting, or military service.[3] Women's neurasthenia, on the other hand, experts blamed on their increasing tendency to be active outside the home; leading physicians prescribed lengthy bed rest to restore their natural passivity. Meanwhile, redoubtable feminist Charlotte Perkins Gilman argued that enforced passivity itself was the root cause of neurasthenia, since it kept women from much-needed engagement with social and political life.[4]

Each of these solutions interpreted neurasthenia as a gender crisis among elites, to be solved by a return to "natural" roles for elite men and women. By appealing to the medical logic of neurasthenia, Roosevelt, Gilman, and others helped define what was distinctive about the elite psyche while also advocating "therapeutic" political action to stave off their extinction. The illness thus helped explain and justify a wide range of supposedly gender-strengthening pursuits in the late nineteenth century, from competitive sports to ecstatic religion, settlement house work to imperialism in the Philippines.

Neurasthenia faded away in the early twentieth century, in part because its increasing celebrity weakened its association with the elite and undermined the cultural logic that had made it so useful. While other labels took its place (nervous breakdown, nervous asthenia, etc.), neurasthenia's decline also reflected real changes in the American mental health professions. Maturing through their experiences of World War I and especially World War II, psychologists and psychiatrists began to eye the great masses of populations beyond the traditional patient pools of asylum residents and nervous elites. Leading lights of the professions recast themselves as advocates of public health, calling for "community mental health centers" to provide preventive attention to early stirrings of mental and emotional problems and tackling broad social problems like racism, alcoholism, marital stress, and others that, they thought, could be explained (and combated) in medical terms.[5]

In keeping with this larger project, after World War II mental health researchers undertook a flurry of major studies and claimed to find an epidemic of stress and anxiety affecting virtually every corner of American society. By their lights, the nation was indeed in the midst of an "Age of Anxiety," as W. H. Auden had warned in his famed 1947 poem. One prominent study, for example, claimed that as many as half of New York City's residents suffered from clinically significant anxiety or related ills, and a host of other researchers reached similar conclusions about other cities. Reflecting psychiatry's new idealistic social vision, such studies often noted pointedly that distress was most acute where poverty or other hardships like racism added to life's challenges.[6]

Mental health professionals' self-proclaimed new social mission, and the new epidemiological findings that underlay it, would seem to have laid to rest the ghosts of elite neurasthenia. But the old nervous illness paradigm proved hardy even amid the parade of theories addressing racism, poverty, and other problems of socially marginal groups. In the torrent of popular attention to anxiety that characterized the postwar decade, physicians and many other public figures continued to associate anxiety with the supposedly complex, intelligent, and self-aware minds and hearts of the affluent.

A classic 1952 study of social class and mental illness by psychiatrists

August Hollingshead and Frederick Redlich revealed just how tenacious such assumptions could be. The researchers found that almost everyone suffered from neurotic anxiety, regardless of class status. But the vast majority of city dwellers who actually sought treatment for their problems were the wealthy, who, the authors hypothesized, had been trained to accept the idea of neuroses and were accustomed to being served by social institutions. They therefore saw little amiss in taking even minor emotional problems to a physician. Their poorer neighbors, on the other hand, skeptical of nervous illness and distrustful of authorities, tended to avoid physicians until their problems progressed beyond neurosis. That explained the overrepresentation of the poor in treatment for schizophrenia at mental hospitals, an endpoint typically arrived at through the criminal justice system rather than the family doctor's office.[7]

This was a powerful indictment of economic injustice in American mental health, and Hollingshead and Redlich's stated goal was to debunk what they saw as the myths that shaped such inequitable social patterns. It is suggestive of the subtle and pervasive power of those myths, however, that even with such admirable intentions the two researchers were ultimately unable to break entirely free of them. Summing up their discoveries at the end of the book, Hollingshead and Redlich unexpectedly abandoned their cogent critique of city dwellers' class-determined "path to treatment" in favor of a different scheme identifying "class-typed" neuroses. A poor neurotic, they argued, expressed his or her anxiety in "antisocial and immaturity reactions" (i.e., by "behaving badly"), while an affluent neurotic was "dissatisfied with himself." In this categorization—ironically, one of the most quoted sections of the study—higher social status meant mental health problems related to self-awareness and complex psychological interiority, while lower social status meant problems manifested through the body, through either physical symptoms or bad behavior. As described by one major psychiatry textbook, the study showed that "psychoneuroses occur more often among people in the upper and middle socioeconomic groups, and the psychoses more often in people from the lower socioeconomic portions of our society."[8]

Another prominent postwar student of anxiety, psychologist Rollo May, made even clearer and less apologetic connections between neurosis and affluence in his ambitious 1950 treatise *The Meaning of Anxiety*. A magisterial survey of psychological thought, May's book concluded with a small experiment of his own devising, in which he assessed the anxiety of ten white and black unwed mothers. Only the middle-class whites, he found, suffered psychologically from their condition; the others were (to his mind) strangely undisturbed. Reasoning that anxiety resulted from a cleavage between expectations and reality, he argued that the middle-class white women's greater distress reflected their higher life expectations. Broadening this argument, May speculated that "individual competitive ambition, chiefly a middle class trait, is intimately associated with contemporaneous anxiety in our culture." Considered alongside his study of psychoanalysts, existentialists, and other schools of psychological thought, May's experiment suggested to him that "there is much *a priori* reason, and some *a posteriori* data, for the contention that neurotic anxiety is especially a middle-class phenomenon in our culture."[9]

Such medical logic also remained commonplace in mass-market media. *Newsweek*, for example, surveying "Tension and the Nerves of the Nation" in 1956, described anxiety as striking a large cast of stock white-collar characters: "the gray flannel suit men . . . the Wall Street tollers of the ticker tape, the doctors, the lawyers, or busy mothers in the home." In the waiting-room magazine *Today's Health*, an article described an experiment in which "executive monkeys" received electric currents to their brain-stem centers, causing them stress and ultimately ulcers. Such emotional stresses, the article concluded, "ulcerate executives and others susceptible to the disease." On the other hand, "non-executive monkeys never develop ulcers." Bestselling anthropologist Margaret Mead described Americans' anxiety in the *New York Times Magazine* as "a large advance over savage and peasant cultures" whose constant fear of starvation and natural disasters gave them little time to develop existential angst. A special report on "The Anatomy of Angst" in *Time*, meanwhile, noted that "what passes for anxiety in the U.S. is really the stress of effort in a land of ambition, competition and challenge." Only as a group became acculturated to America did it begin to share in this national trait. African Americans, according to the article, were currently experiencing more anxiety as their opportunities for social advancement expanded.

"Puerto Rican and Mexican immigrants," on the other hand, "will have their innings with anxiety later."[10]

No seminal figure like George Beard emerged to formalize these often casual theories that used anxiety as a medical shorthand for race and class differences. But in a way this only allowed the reasoning linking the "age of affluence" to the "age of anxiety" to circulate even more widely, because virtually anyone could claim to be an authority on it. Psychoanalysts meant something quite different by "anxiety" than psychologists did when they spoke of "stress" or "tension," and popular references to "worry" and "nervousness" only further clouded the picture—but all made some claim to the mantle of medical or scientific authority. And cutting through this theoretical heterodoxy ran a consistent thread: all of these various stories were about affluent white Americans, and all saw anxiety as signaling their psychological depth and complexity.[11]

As with neurasthenia, the apparent epidemic of anxiety also allowed cultural crusaders to diagnose problems in the white-collar classes and advocate therapeutic social and political solutions. Not since the era of Theodore Roosevelt had there been such widespread and harshly self-critical assessments of the state of mainstream white culture. A virtual cottage industry sprang up purveying withering portraits of anxiously conformist "grey flannel suit" men, their neurotically overbearing wives, and their too-soft children. And the cause of all this debilitating anxiety, according to a wide range of critics, was a brewing crisis in gender roles. White-collar men were losing their authentic, tough masculine vigor, while their wives had strayed from the path of domestic femininity. Such fears were standard fare in postwar campaigns to reestablish male authority at home and at work after the disruptions of the Great Depression and the war. For psychiatrists, politicians, scholars, and many others, anxiety provided a medical logic supporting this broader social agenda.[12]

Sociologist David Reisman's influential bestseller *The Lonely Crowd*, for example, painted a sorry portrait of white-collar men as "other-directed" souls driven by a constant, free-floating anxiety to please those around them. Subsumed into the corporate machine, they had lost the virile autonomy of their "inner-directed" forebears, who in earlier and simpler times had followed their own inner moral compass—without the anxiety. Self-proclaimed literary rebel Norman Mailer, meanwhile, famously advised American men to conquer their "collective failure of nerve" by becoming "White Negroes," that is, by emulating the anxiety-free, instinctive pleasure seeking of black hipsters.[13] The echoes of Margaret Mead's "savage and peasant cultures" could not be clearer in either case. Anxiety might be a sign of advancement and intelligence, but it was also a weakness that undermined a supposedly more authentic and straightforward masculinity.

Such rhetoric also surfaced with great significance in the political arena, in the pivotal struggle between the anticommunist right and newly hawkish Cold War liberals. Both sides appeared to agree on the enemy: an overrefined, fearful political elite incapable of taking a tough stance against the real-world evil of the Soviets. For Arthur Schlesinger, historian and key political adviser to postwar Democrats, frightening global developments had made anxiety "the official emotion of our time," driving many of the weak-willed to embrace totalitarianism on the left or on the right. Schlesinger's influential book *The Vital Center* advocated a militaristic, manly liberalism as the best way to rescue a nation "consumed by anxiety and fear"—a situation he tellingly described elsewhere as a "Crisis of American Masculinity."[14] On the political right, Senator Joe McCarthy and others openly linked an effete "softness" on communism to homosexuality and suggested that liberal elites in the State Department were so far gone in this direction that they had essentially ceased to be men at all.[15]

Anxiety was also a watchword among the powerful voices calling on women to devote themselves fully to housewifely duties. Surrounded by a multitude of newly available roles, wrote the authors of the archetypal 1947 book *Modern Woman: The Lost Sex,* "modern woman is a bundle of anxieties." Only a focus on their female essence, and a return to traditional roles, would permit women to find "solid satisfaction, peace of mind, [and an] easing of deep inner tensions."[16] Freudian professional psychiatry officially agreed, labeling as neurotic or even psychotic women who deviated from submissive domestic roles—especially middle-class stay-at-home housewives who devoted too much "anxious mothering" at their children.[17]

Like neurasthenia, then, anxiety proved useful for a wide range of white and white-collar cultural crusaders. On the one hand, its supposed preference for the nation's upper ranks offered a medical shorthand for their self-awareness, intelligence, and psychological complexity. It also allowed them to take on a "sick role" and get treatment rather than be punished for bad behavior or institutionalized for insanity. On the other hand, the ubiquity of anxiety suggested that white-collar America was endangered and in need of rescue. As in the late nineteenth century, the most visible of these therapeutic rescue campaigns hinged on describing anxiety as a gender crisis: the comfortable classes were becoming sick because modern living squelched their true masculinity and femininity—older, less civilized but more vital ways of being male and female. To cure the illness meant returning to "traditional" gender roles: men to strength and authority at home, at work, and in politics, and women to domestic submissiveness and housewifely service to husband and children.[18]

All of this is to suggest that anxiety became the signature illness of postwar culture—the "age of anxiety"—in part because of its usefulness in an era of intense gender politicking. Certainly the newfound power of Freudian psychiatry helped place anxiety in the public eye, but it was the continuing logic of the neurasthenic tradition that made it central to so many postwar cultural campaigns. It was these gender politics that Miltown's advertisers confronted in marketing their new wonder drug. Later observers have tended to assume that the central drama of the psychotropic revolution was its challenge to Freud. At least initially, however, advertisers did a much better job soothing relations with Freudians than with champions of anxiety as a crisis in gender.

"Happy Pills" and the Age of Antianxiety

The neurasthenic paradigm rested on certain ideas of anxiety: that it was a characteristic illness of affluence and intelligence, and that the best treatment was stronger adherence to normative gender roles. For those seeking to reassert male authority after the dislocations of depression and war, anxiety thus served as a disciplinary illness, creating new cultural campaigns and medical practices around the argument that straying from natural, authentic masculinity or femininity could literally make you sick.

These cultural campaigns forged "common sense" about emotional illness that Miltown's supporters—and detractors—worked with as they struggled over the role of tranquilizers in American society. Thus, for example, virtually everyone agreed that tranquilizers were drugs of and for the middle classes, because anxiety was an illness of affluence. In this regard the blockbuster drug phenomenon expanded the reach of neurasthenic logic, adding powerful new voices like advertisers eager to spread the gospel of anxiety. But advertisers also introduced changes to the nervous illness story. In their quest for sales they sought to broaden the kinds of anxiety that counted as medical, for example, which pushed against the social exclusivity of the neurasthenic model. Their enthusiasm for tranquilizing men as well as women also ran afoul of neurasthenia's gender politics. They were, in short, working within the logic of the nervous illnesses but also trying to transform it by establishing tranquilizers as consumer goods available to everyone. The result was a dynamic dialogue between gender critics, advertisers, and physicians over the lessons about selfhood and psyche that Miltown and its cousins supposedly had to teach. This dialogue, I argue, ultimately helped stamp the infant discipline of biological psychiatry with the gender politics of the 1950s.

To understand these battles over Miltown, we need to start by examining how physicians, advertisers, and journalists worked together to position the tranquilizers as middle-class drugs.

In professional medical literature, physicians consistently described the anxious patients treated with tranquilizers as white-collar "brainworkers" or their wives—people whose psychological complexity and intelligence put them at risk. One group of Miltown-advocating psychiatrists, for example, described the "tension states" they treated as "characteristic of a group of individuals who are, by reason of their inherent qualities and training, the finest product of our culture," people whose "fundamental drives and abilities . . . serve to make them basic leaders [and] predispose them to accomplishment." The kind of anxiety they saw as calling for Miltown was thus "so common as to be statistically 'normal' among professional persons," but "rare among Southern Negroes or reservation Indians."[19]

Leading psychopharmacologist Karl Rickels made a similar point in his address at a 1964 symposium (funded in part by Carter Products) celebrating Miltown's first decade. Minor tranquilizers, he proposed, were particularly appropriate for "middle to upper socio-economic class patients, who are more intelligent, have higher drive levels and ambition, and are less compliant and dependent." Simple sedatives, on the other hand, would suffice for "lower socio-economic patients with low drive levels, low ambition, highly compliant behavior, and a strong passive dependent personality structure." He concluded by noting (apparently without irony) the convenience of this "interesting sociological phenomenon": "Patients who cannot afford to pay for the more expensive tranquilizers are the same patients who are least affected by sedation and, therefore, may benefit most from the less expensive barbiturates."[20]

While such direct references to social status were relatively rare, they nonetheless stood out by virtue of the nearly complete absence of counterexamples. In the extensive literature about the tranquilizers, virtually no physician or other observer broached the subject of Miltown's possible value to the ranks of the nonwhite or poor.[21] Moreover, the cast of characters peopling the medical literature on the minor tranquilizers presented a remarkably homogenous face. They were, for example, a tense salesman who had "driven himself to the top of his company's sales force," or a "business executive" whose company had fallen upon hard times. They were ambitious lawyers or overtaxed middle-class housewives. A careful search of early literature does turn up a panic-stricken electrician, appearing in a textbook on psychiatric drug treatment. Even this relatively respectable workingman was breathing in rarified air, however: the other five cases with whom he appeared included a stockbroker suffering from "uncertainty, apprehension, and phobias," a physician with an inferiority complex, an anxious engineer who had taken to drink, a graduate student with family and school troubles, and a housewife of at least some means grappling with a husband and prickly in-laws.[22]

Advertisers worked with these medical assumptions to make their case for the importance of tranquilizers. Despite insisting that anxiety was a universal human trait, for example, ad campaigns invariably centered around white characters inhabiting the environs—and facing the challenges—of America's comfortable classes. Of the thousands of ads in the *Journal of the American Medical Association* and the *American Journal of Psychiatry* through the mid-1970s, for example, only two depicted a nonwhite model—and one of those was a man whose "everyday" anxiety did *not* merit treatment with tranquilizers.[23] Moreover, in this lily-white world men wore coats and ties in business settings, and women were surrounded by similar indications of privilege such as new household appliances, pearl necklaces, and stylish clothes. One 1959 ad, for the amphetamine-barbiturate combination Dexamyl (marketed as a tranquilizer-antidepressant), went so far as to describe a "businessman's syndrome . . . marked by nervousness, worry, tension, and feelings of futility."[24] One looks in vain for a marketing campaign centered around, say, a plumber too anxious to pick up a wrench.

This tendency to portray Miltown as a white-collar drug bore a complex relation to reality. As Karl Rickels had noted, Miltown and other new tranquilizers were indeed expensive. According to *American Druggist*, tranquilizers, at $4 to $6 for a two-week supply, cost twice the average price for all drugs and were the most expensive class of medicines available between 1955 and 1964. Getting a prescription, moreover, meant seeing a doctor—an economic and cultural obstacle for many outside the middle classes.[25] This combination of narrow access and high cost may have helped skew use patterns by class and race: according to most utilization surveys in the late 1960s and 1970s, wealthy professional whites used the drugs at higher rates than working-class whites and nonwhites.[26] Yet even these studies made one thing absolutely clear. Owing to the large numbers involved, a great many lower-income whites and African Americans were taking Miltown despite the obstacles. They just failed to capture the same level of attention and interest as their better-heeled peers.

Miltown's supporters, then, helped advertise the idea of anxiety as a key component of white-collar life, in keeping with the neurasthenic tradition. But in the process of making Miltown into a blockbuster celebrity drug, the multilayered forces of commercial medicine significantly altered the storyline of what anxiety and its treatment meant for America's white middle classes. Their effort to broaden the kinds of anxiety that merited medical treatment, for example, tended to push against the social exclusivity of the neurasthenic model. Virtually any unpleasant experience suffered by the well-dressed but miserable population inhabiting the ads seemed to have a legitimate medical component—and the "normal problems" of "everyday patients," or "that blah feeling," hardly suggested profound existential states. Moreover, the consumerist narrative of anxiety resolved through use of a simple, purchasable pill con trasted sharply with the notion of anxiety as a social illness requiring political and cultural change. And, as noted in chapter 1, such messages were designed to ricochet through popular as well as medical arenas—magazines, gossip columns, radio and television, and so forth.

In some ways this more intense push for therapeutic territory mirrored broader trends in the rise of "therapeutic culture" in the 1950s, as mental health professionals sought to claim authority over more and more realms of life experiences.[27] But the Miltown phenomenon differed in its marrying of open consumerism to its therapeutic promises. In fact, the marketing campaigns were so enthusiastic, and so unlike traditions of prescription drug advertising, that minor tranquilizers became famous as products as much as medicines. Beyond medical and popular media, for example, Miltown and its competitors earned fame in financial periodicals like *Barron's, Financial World,* the *Magazine of Wall Street,* and *Advertising Age,* which celebrated their miraculous profit-making power even during the late-1950s recession.[28]

Even outside these business-oriented publications, the buzz of promotion and profit surrounded the new wonder drugs almost as densely as their therapeutic promise. Indeed, based on the first rash of popular media coverage, most Americans could have been forgiven for assuming that Miltown and other "happy pills" were fashionable new consumer goods, available to anyone who wanted them. As early as 1956, *Time* magazine described a Miltown "craze" in Hollywood, where "the craving for the 'don't-give-a-damn' pills" left drugstores with backlogs of unfilled orders. "I wish the government would subsidize slot machines for tranquilizers on every corner," the magazine quoted a Beverly Hills psychiatrist as saying.[29] A *New Yorker* cartoon by Charles Addams, reproduced in *Life* magazine, showed a businessman in the subway passing what looks like a candy machine but is actually a tranquilizer dispenser. "Get Thru The Day!" it advertised cheerily.[30] *Newsweek,* meanwhile, reported that the pill's popularity had led to drugstore shortages and ran a picture of a store window posted with the sign "Yes—We Have MILTOWN *and* EQUANIL.*" The author concluded, "Over-the-counter sales of these prescription drugs have become inevitable" owing to their popularity.[31] *Consumer Reports* ran an article guiding its readers to make intelligent choices about the drugs.[32] *Life* magazine detailed the evolution of a pill, including extensive market consultation on what color it should be and

In Los Angeles the fad spreads

Selling "wonder drugs" in the postwar consumer culture. *Source: Newsweek* photo, May 21, 1956, 70; Charles Addams cartoon printed in the *New Yorker* and *Life*, 1956, © Tee and Charles Addams Foundation.

other consumerist angles.[33] *Fortune* magazine announced that the nation was on the verge of a "choose-your-mood society" that would allow "essentially normal people" to overcome "'infantile colic, senile dementia, and almost every emotional complaint that mortal man has ever been able to jam in between.'"[34]

Miltown and the Postwar Masculinity Crisis

Later in the century feminists would critique commercial medicine for, among other things, pushing drugs like Valium on women whose unhappiness was not medical but political. And indeed, women used tranquilizers and antidepressants at twice the rate of men. But before they became famed as "women's drugs," the tranquilizers first attracted widespread public criticism in the 1950s because of the supposed threat they posed to the already beleaguered masculinity of America's white-collar classes. Drug marketers' hybrid medical consumerist model clashed in important ways with the notion of anxiety as a social illness best cured through a return to proper gender roles—especially for men. According to the neurasthenic tradition, easy access to tranquilizers would be the worst possible thing for men, who if anything needed to be energized rather than further relaxed. Tranquilizers smacked of bed rest, the quintessential cure for women who needed to be returned to their "natural" passivity.

The initial marketing-powered publicity surrounding Miltown, however, made few gender distinctions: the new wonder drug was for everyone. Physicians, advertisers, and journalists wrote in neutral terms of "everyday patients," the "need for equanimity," or just plain "anxiety." The only image in Miltown's initial advertisements was of the "Miltown molecule," and most other early tranquilizer ads also mixed universal images (a calm beach, for example) with technical language describing clinical trials (see chap. 1). The first round of popular reporting also spoke in relatively generic terms about the "craze."

To those concerned about the midcentury "crisis of masculinity," Miltown was a perfect symbol of the forces luring middle-class men into a comfortable and conformist passivity. Already in the public eye thanks to the marketing machinery of commercial medicine, the ubiquitous wonder drug provided an opportunity for these gender critics to dramatize their arguments about failed masculinity. In the popular media, amused and humorous reports of "happy pills" were quickly accompanied by darker warnings about what *Business Week* called the coming "tranquil extinction."[35] As one psychiatrist told *Life* magazine, "I don't look with any favor on a society where everybody just floats around in his own tub of butter. A certain amount of tension and alertness is essential to keep things straight in life."[36] What needed "keeping straight," apparently, was the middle-class male psyche. Reacting against advertisers' visions, gender critics now portrayed tranquilizers as a consumerist threat and anxiety as the essence of the imperiled masculinity they were trying to preserve. The logic had shifted, but the result was the same: a wave of popular jeremiads stressing the importance of "true" masculinity to America's medical and political health.

The white-collar workplace was a key arena for Miltown's critics—not surprising, given the amount of ink expended on the "organization man" and his discontents in the 1950s. *Look* magazine's 1956 survey of mind drugs, for example, recounted the story of a junior executive whose wife "slipped him some tranquilizing pills she had obtained from a friend" to ease his anxiety about an upcoming speech:

> They worked like a charm: his worries just melted away. When the big moment arrived, the speech was still unprepared, but the junior exec faced his audience unconcernedly and delivered his talk off the cuff. It was a complete flop, but the young man, unaware that he had made a fool of himself, was grinning happily when he sat down.

Women's Home Companion offered a similar tidbit about a business executive whose justified worry about an impending financial crisis was too thoroughly allayed by tranquilizers: "The crisis struck and ruined his business, as he looked on with apathy." Robert Felix, director of the National Institute of Mental Health and no enemy of new drug treatments, nonetheless aired doubts to an interviewer for *U.S. News and World Report* about "the executive who [takes tranquilizers] to tide himself over a crisis that his predecessors sweat through." The *New Republic* began an editorial by describing a cartoon about a wife who offers this advice to her husband: "You skip your tranquilizer. Watch for your boss to take his. Then hit him for a raise." "The remark was in a cartoon," the article continued, "but it's no joke."[37]

For the many voices spreading worry about failing masculinity, Miltown symbolized more than workplace passivity. It also played a key role in a grander drama: the rise and fall of Western civilization itself. In the neurasthenic paradigm, anxiety was the distinguished problem of an "advanced" civilization. But anxiety was a double-edged sword, indicating superior culture while also posing a threat to that culture. Responding to that threat with tranquilizers instead of a return to masculine authenticity only made the problem worse. *Life* magazine, for example, noted that tranquilizer takers "have reported that *larger* efforts, or *larger* thoughts seem to escape them; somewhere, a drive to cut through the tougher difficulties of a situation has been reduced." The editors of *Business Week*, warning about the threat of "tranquil extinction," argued with rare poetic passion that a nation of pill-poppers "would mean that man had given up the effort to create beauty and love out of the fierce tensions to which his nature exposes him." *American Mercury* chimed in, attributing to "that wonderful frustrated feeling" every advance in civilization and technology and asking why man should "cheat himself and posterity" by hampering the development of the already too rare "nervous genius of man."[38]

Similar stories about Miltown also surfaced with some regularity in medical literature. "Our civilization has been built on the divine discontent of tense men," wrote one physician. "Perhaps Columbus could have discovered the New World while taking tranquilizers, and Beethoven might have been able to compose his symphonies, but I submit that if they had been full of [Miltown] they wouldn't have bothered." Reasoned another, "The desire to improve himself, to earn a better living, to make bounteous provision for his family and to enjoy a higher station in life is generated by tension, which lights the fire of ambition." *JAMA* itself published the concerns of two psychiatrists who feared that "we are to be completely tranquil under all circumstances and let someone else 'do the worrying,' try to get ahead, or be successful."[39] The president of the American Medical Women's Association feared that tranquilizers could "contribute to the regression of a civilization that has made some of its greatest advances in response to stress." A speaker at the New York Academy of Medicine warned his audience that anxiety was crucial for scientific advancement and concluded that "the employment of tranquilizing agents is loaded with potential doom for the race as the highest of evolved creatures."[40]

These decline-of-Western-civilization arguments took on even more compelling urgency when put in the context of Cold War anticommunism, which was rife with fear of male "softness" and homosexuality. Novelist Aldous Huxley, normally a reliable pro-drug voice in the early psychotropic era, challenged the readers of the *Saturday Evening Post* with a dire scenario—"not, alas, a hypothetical case," he pointed out—of two competing societies. In one, tranquilizers are available but expensive, meaning that they are restricted to elites, who happily down billions of the "complacency-producing" pills yearly. In the other, tranquilizers are rarer, and elites "do not resort, on the slightest provocation, to the chemical control of what may be necessary and productive tension." Which of these societies, he pointedly asked, "is likely to win the race?"[41] The editors of New Jersey's state medical journal agreed: "It may be that anxiety is the divine spark that keeps us going . . . a culture that has been soothed by solacing syrup of tranquilizers could fall prey to a civilization that is dynamically powered by its own discontent."[42] Ultimately, opined the *Nation*,

> tranquilizers can affect the whole society—its cultural dynamics, stasis, decay. What happens to a people accustomed to avoid all anxiety, who know no fear, who need not think, or love or hate? What happens to a people without an urge to create? Or to people who must take the proper portion before being able to do any of these? And what happens to such people when they meet up with "normal" people, or with "noble savages" who have not had the advantages of the most advanced psychopharmacology? . . . As we watch over the decline of the West, we see the beams—the bombs and the missiles; but perhaps we miss the motes—the pretty little pills.[43]

In suggesting that the fate of Western civilization hung on the edge of a pill bottle, the *Nation* joined a host of gender critics who helped push Miltown into even greater heights of celebrity by attacking it. These latter-day neurasthenic crusaders exploited the new spaces of commercial medicine to spread their message about the importance of an essential, natural masculinity. In doing so they also changed key elements of their story: anxiety still figured as a mark of advanced, ambitious intelligence, but it was now also portrayed as the essence of authentic masculinity, a vital resource for remaining dissatisfied amid the blandishments of an increasingly soft consumer culture. Tranquilizers were no cure; they were the essence of the problem, undermining American manhood with the pap of comfort. "The characteristic drug of our age is the tranquilizer," Arthur Schlesinger lamented in *The Politics of Hope.*[44] A true man, these critics suggested, possessed a complex and paradoxical character that balanced animal strength with civilized intelligence. He was at odds with himself, marshaling his deep primitive energies for the advancement of civilization. In effect, he combined the best of both worlds, with the vitality of the rude and barbarous driving the controlled refinement and brilliance of the "highest of evolved creatures." The psychic cost of this inner conflict was anxiety, now portrayed as unassailable proof of masculine toughness as well as psychological complexity.

However far afield such stories about tranquilizers might have strayed from the experience of actually using the medicine, their high visibility and wide circulation in both popular and medical media helped shape commonsense understandings of Miltown in the late 1950s. The notion of tranquilizers as a consumerist threat to masculinity did not come entirely from any one place; it emerged from a dynamic interaction among physicians, advertisers, patients, and gender critics. Through clinical trials and prescriptions, physicians and patients helped convey the effectiveness and power of the drugs. Advertisers added their own slant to the drug's capabilities, integrating prescription tranquilizers into the simpler before-and-after narratives of the consumer culture. All three groups helped establish Miltown as an object of popular fascination, particularly among the white-collar classes. Once the drug had attracted public notice, other voices joined in maintaining its celebrity: the cultural critics who found it a useful symbol for middle-class masculinity in decline.

Advertising Authentic Men

The Miltown media frenzy showed just how thoroughly commercialization had changed the production and circulation of medical knowledge in America. Thanks largely to marketers' efforts, blockbuster drugs now commanded public attention through reporting in the health sections of popular magazines, celebrity gossip columns, consumer or lifestyle pages, business and economy sections, and so on. These cultural spaces became the lifeblood of commercial medicine—a system for reaching broad audiences with a consumerist slant on new technologies. But as a public forum it also invited others to weigh in, too: journalists, celebrities, cultural crusaders, political activists, and others who populated the world of mass-market media. These groups joined physicians and advertisers in helping produce popular knowledge about tranquilizers and what they portended for American society.

By the early 1960s, for example, the outcry against the tranquilization of American men was working with other factors to make Miltown and its cousins seem inappropriate or even dangerous for men. These attitudes were reflected in the first studies of tranquilizer use, which showed that men were prescribed the pills only half as often as women. The use of Miltown as a symbol of masculine decline was so successful, in fact, that by the end of the decade minor tranquilizers had simply ceased to be an issue of public relevance for men; Librium and Valium, Miltown's successor drugs, would be known almost exclusively as women's drugs and would be discussed in very different contexts.

Despite this dramatic shift in tranquilizers' popular reputation, however, drug advertisers were unwilling to give up on the still sizable male market. They responded to popular criticisms with new ad campaigns, suffused with explicitly gendered images and language that portrayed tranquilizers as aids in restoring proper masculinity as well as femininity. Such ads maintained and even helped expand the number of male drug users, even as their proportion in relation to women remained consistent. More important, these ads introduced 1950s-style gender conventions into the slowly emerging discourses of biological psychiatry. In essence, marketers helped interpret the science of the brain in ways that supported the era's white-collar masculine ideal. Here, hidden behind the slowly building furor over the "tranquilization" of women, advertisers participated in the equally important but far less examined cultural project of building biological explanations for particular concepts of masculinity.

The first signs of these efforts began appearing in the early 1960s, when advertisements began to tell new kinds of stories about the value of tranquilizers and other drugs for white-collar men. Most obvious, they began to emphasize how the drugs would return these men to properly masculine vigor, decisiveness, and achievement in office settings—a more gender-specific message than the "everyday patient" theme of early ads and a clear rebuttal to popular fears of workplace passivity. A parade of men in business attire marched alongside text promising drugs that seemed to respond directly to popular worries. Marketers for new contender Striatran from Merck, for example, claimed in 1960 that their drug produced "*alert* tranquility." The image showed a productive-looking businessman whose desk has been transported to a beautiful outdoor scene.[45] Librium's advertisers addressed the subject head-on in 1962: "In many situations," they admitted, "a certain level of anxiety is normal, even desirable." But "as its weight increases" it can "slow down efficiency" or even bring it to a standstill. The image was of a man's face—his collar and suit jacket just visible—jarred by stress lines and overlaid by a ruler.[46] Valium's promoters weighed in a few years later, admitting in a *JAMA* ad that "psychic tension is part of living—a useful part, to some degree, since it engenders drive, aspiration, creativity. Emotional tensions 'make the world go 'round'—but for many people, they sometimes make it whirl at too dizzying a pace."[47]

Advertisers did more than promise to rescue masculinity imperiled. They also circulated new ideas about how and why tranquilizers worked, ideas that provided scientific backing for the 1950s masculine ideal. These new ideas came out in a diverse set of campaigns during the 1960s and into the 1970s that focused not on Freud-inspired visions of psychic anxiety but on the biochemical machinery of the body.

One simple manifestation of this was the multitude of ads featuring images of men (and sometimes women) dwarfed by their own enormous, glowing, or spikily irritated stomachs, intestinal tracts, or hearts. These ads were typically for drugs that combined a tranquilizer and another medicine, and which could therefore be sold as cures for ulcers, hypertension, or some other "executive-type" illness without mentioning anxiety. A similar genre of ads drew parallels between mechanical and mental situations, such as a Valium ad with an image of stress patterns on a piece of plastic "under tension," or a Librium ad displaying a person's head as composed of springs, switches, and other pieces of machinery.[48] Like the internal-organ ads, these images displaced the arena of action from the mind to the body, avoiding any dangerous suggestion that a man's mind might be pacified while still recommending tranquilizers for common conditions such as stomach discomfort.[49] A related set of Valium advertisements promoted the drug as a muscle relaxant for sports injuries, again highlighting masculine bodily problems.[50]

Other advertising campaigns took such logic a step further, suggesting that the mind itself was part of the body's animal machinery. In these campaigns the male psyche (never a woman's psyche) was depicted as being made up of two essential parts: a primitive or animal self (the hindbrain), and a modern or civilized self (the forebrain). Anxiety resulted from the conflicts between these two physiological selves.[51] Men constantly battled their primitive animal selves, which showed no signs of being worrisomely quiescent. Quite the opposite: the animalist parts of men were too vigorous to ever be fully contained. The stress of trying to keep them in check, in fact, was the cause of anxiety. Anxiety was thus "rescued" as a sign of problematic animal strength, even while tranquilizers were still shown to be necessary to harness that strength.

These animalist campaigns were inaugurated by Librium's famed "new way to calm a cat" series in 1960, which touted the drug's ability to tame wild animals. "Characteristically vicious macaque monkeys," rats, and mice all became "manageable and friendly" under Librium, while still offering a "striking contrast to 'doped-up'" animals under stronger tranquilization—they were, in short, "calm but alert." In the same journal, Librium was advertised for a very different animal, a publishing-house "trouble shooter who had troubles of his own." Endless deadlines had left the man in "a continuous state of tension," resulting in nearly two decades of recurring ulcers.[52] This man was a classic neurasthenic type, a white-collar brainworker suffering from the hurly-burly of the competitive marketplace. And yet the implicit parallel between him and the wild animals in the other Librium advertisement suggested that it was his own aggressiveness, not any passive overcivilization, that had produced his anxiety. (The ideas in the ad were also helpfully circulated by an industry-friendly journalist in *Today's Health* magazine, which reported that "executive monkeys" receiving electric currents to their brain-stem centers eventually get ulcers, while "non-executive monkeys never develop ulcers.")[53]

A 1969 Librium advertisement illustrates this kind of approach perfectly. The headline text ran, "Man's ancient heritage of neurohormonal defenses subverted by chronic anxiety." Beneath ran a picture of a caveman and a businessman in parallel defensive postures, each responding to an external threat. The caveman's response was "appropriate," allowing "primitive man to meet short-term threats [such as the pictured tiger] effectively." The same reactions in the businessman, however, were "inappropriate" to the "complex, long-term psychic threats" of the modern world. The explanation is worth quoting at length:

> Modern man is equipped with the neurohormonal physiology and responses of his prehistoric ancestors appropriate for physical emergencies. Today, however, threats to security occur predominantly in the mental and emotional spheres . . . They may tend to persist indefinitely with the result that various organ systems remain chronically and inappropriately mobilized . . . [This can] lead to permanent structural changes in various bodily systems, resulting in organic disease.[54]

Here was a remarkable reworking of the neurasthenic paradigm. Instead of interpreting anxiety as evidence of weakness and overcivilization, these and other ads in the 1960s identified it as a sign of strength—an indication of the powerful inner caveman.

Valium's advertisers went a similar route in 1968 with a multipage appeal in *JAMA* purporting to explore "anxiety and the male psyche." Among other things, the text warned that higher-income groups suffered anxiety the most. Once again, however, the explanation lay in men's strength rather than their weakness. The attainment of their high status had taken "fierce, hard climbing." This climbing, in turn, reflected innate biological drives: "In men there may be an androgen-mediated-phylogenetic vestige . . . of the mechanisms seen more highly developed in some others of the mammalian species, for the claim and defense of territory and breeding-ground rights." Put simply, men's primeval urge to claim territory and mates was written into their very biology, driving the "best" ones inexorably onward past sustainable levels. "Clearly," the text concluded, "the masculine man—with his aggressive, hard-driving and fiercely competitive nature—will be with us for some time to come. Which, despite its attendant problems, is probably all for the best."[55]

In these campaigns advertisers reworked the story of anxiety to re-

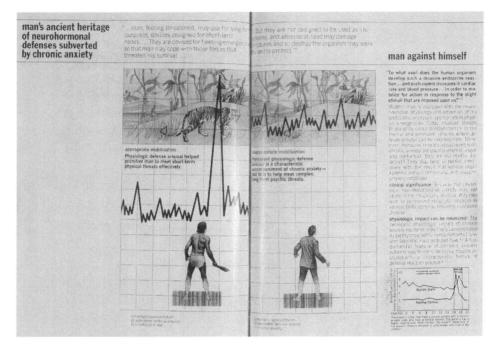

Tranquilizers help tap into the caveman within? *Source:* Librium ad, *JAMA,* September 15, 1969. Courtesy of the New York Academy of Medicine Library.

cover it as an indicator of strength while retaining an important role for tranquilizers. In essence, America's "best" men were too strong, too masculine, and were failing in their efforts to harness that strength—to control themselves. Tranquilizers could help by calming the restless animal within, keeping it alert and active but making it less likely to fight, to react unreasonably to stimuli, and pointlessly waste psychic energy. In the hands of advertisers, then, tranquilizers had evolved from a looming threat to American manhood to the ideal solution to such threats. In these marketing campaigns, even the most civilized men had a beast within, ineradicable and pure. The stronger the beast was—and the stronger the intelligence devoted to harnessing it—the more likely debilitating anxiety could result. Tranquilizers and other drugs could help manage this delicate but crucial process, thus rescuing American men from themselves. It was no accident that this psychic narrative emulated almost precisely the vision of manhood put forward by Cold War masculinists, for it had been designed, at least in part, in response to just such visions as expressed in popular fears of Miltown.

Tranquilizer advertisers were not the only ones beginning to see connections between humans and animals in the 1960s. The later years of the decade saw the rise of "human zoology," or "sociobiology," as it would later be dubbed by eminent biologist Edward O. Wilson. Sociobiologists purported to study society as but the latest phase in the animal evolution of the species, fundamentally shaped by inheritances from long-ago animal and caveman ancestors. In an explicit effort to avoid reprising eugenics and Nazi-era theories about race differences, Wilson and other sociobiologists focused on species-wide characteristics. This had the effect, intentional or not, of zeroing in with new intensity on gender differences, which could be described in universal terms as relating to all humans. The most widely popular sociobiology book, Desmond Morris's mega-bestseller *The Naked Ape* (1967), characterized people as having changed little from their primate ancestors. Men were ranging, aggressive hunters who sought out many sexual partners. Women were biologically predisposed to cling to a single mate while focusing on reproduction and family.[56]

The popularity of sociobiology came about for many different reasons, and psychopharmacology was only one of many strands of scientific logic that appeared to support its obsession with brute animal masculinity. But in the 1960s, psychopharmacology was one of the few sciences that could believably point to concrete, practically useful technologies as evidence of its claims. It was also a science with enormous marketing power both interpreting and disseminating those claims. Moreover, millions of people were personally experiencing the logic of psychopharmacology when they or someone they knew used pills like Valium. This was, in short, a powerful system of distributing sociobiological ideas about the self in direct, even intimate ways to very broad populations.

The irony of this pharmacological construction of virile manhood is that President John F. Kennedy, whom scholars identify as the public figure most carefully crafted to meet the new standards of elite masculinity circulated during the 1950s, maintained his famously vigorous public image only through a strict regimen of psychotropic drugs.[57] That these drugs were mostly amphetamines, rather than tranquilizers, speaks both to the cultural demands on men at the time and also to the tenuous connection to biological reality of the gender claims in the first place.

"The Problem That Has No Name": Tranquilizers, Feminists, and the Neurasthenic Tradition

Busy worrying over the ramifications of tranquilizers for American manhood, few observers in the 1950s and early 1960s seemed to notice that women already took the lion's share of the pills. In fact, according to comprehensive drug utilization surveys begun later in the 1960s, women used tranquilizers at rates twice that of men. True, women used the medical system more than men in general, but this did not explain the disparity. Women were simply more likely than men to leave a doctor's office with a prescription for a tranquilizer—a pattern not replicated for prescriptions in general.[58] Of all the numbers produced by the slew of epidemiological studies over the next decades, the gender difference was unquestionably the most striking and most universally replicated. Tranquilizers were, at least in this important sense, "women's drugs."

Despite this fact, initial public attention virtually ignored women's tranquilizer use. As a commercially ubiquitous "happy pill" for men, Miltown was ideal for drawing attention to a supposed crisis in white-collar masculinity. But the same gender critics who were horrified at the tranquilization of America's "best" men were likely to applaud the phenomenon in women. After all, according to these critics, anxiety, neuroses, and other ailments plagued affluent women because they had abandoned their true domestic natures. To fix these women's problem meant to revive their feminine qualities of submissiveness, nurturance, and familial devotion.[59] No verbal acrobatics were required to explain how tranquilization with Miltown, or its successors, fit with this goal. For the many public figures advocating a return to "traditional" housewifely roles, women's use of the pills seemed natural and unremarkable.

The wonder-drug marketing machine thus worked with, rather than against, proscriptive ideas of femininity—especially after advertisers introduced stronger gender messages into their campaigns in the early 1960s. At this time, as we have seen, advertisements began to portray tranquilizers as tools to help restore vigorous, ambitious masculinity.

They also began to promise that drugs could return women to happiness in housewifely settings, having regained their supposedly natural ability to serve others, particularly their family. A 1960 Meprospan ad, for example, followed a woman who has "enjoyed sustained tranquilization all day" as she "stays calm . . . even under the pressure of busy, crowded supermarket shopping" and "listen[s] carefully to P.T.A. proposals." This ad promoted Meprospan, but in doing so it was also selling the normative housewifely role.[60] In ads for the sedative-stimulant Dexamyl, a smartly dressed man has regained "the ability to make decisions, the emotional energy to complete his work," while a woman shown hanging flowered curtains has rediscovered "an interest in her surroundings, a feeling of well-being."[61]

Some ads were even more explicit in identifying housewifely dissatisfaction itself as an illness. "A lot of little things are wrong," mused one homemaker in a 1970 ad for the tranquilizer-antidepressant Sinequan: "headaches, diarrhea, this rash on my arm. And sometimes I think I don't like being married."[62] Here advertisers "medicalized" marital unhappiness by suggesting that it was merely one of a list of minor physical symptoms accompanying an underlying mental or emotional disorder. Perhaps the most unsettling such ad was in a 1970 pitch for the antidepressant Vivactil, which promised to get "the patient moving" (down the stairs with the laundry) and then to get "her mood improving" (with her alluring expression inviting a sexual interpretation).[63]

If the many proponents of a return to "traditional" gender roles found little to complain about in these appeals, however, new cultural activists —feminists—were beginning to pay closer, and more skeptical, attention. The earliest stirrings of this attention began with Betty Friedan, a pioneering leader of second-wave feminism among white middle-class women. As biographer Daniel Horowitz has noted, Friedan purposefully aimed her message at white-collar housewives, in part to fend off any potential red-baiting that might dog more broad-based organizing. Her pathbreaking bestseller *The Feminine Mystique* (1963) did just that, establishing an imagined community of well-off housewives suffering from the soul-killing limitations of enforced domesticity.[64]

Part of Friedan's success lay in her use of existing neurasthenic rhetoric, retooling it for her own purposes. As we have seen, many powerful cultural voices (including drug advertisers) were already circulating the

The physician listens to a tense, nervous patient discuss her emotional problems. To help her, he prescribes Meprospan (400 mg.), the only continuous-release form of meprobamate.

The patient takes one Meprospan-400 capsule at breakfast. She has been suffering from recurring states of anxiety which have no organic etiology.

She stays calm while on Meprospan, even under the pressure of busy, crowded supermarket shopping. And she is not likely to experience any autonomic side reactions, sleepiness or other discomfort.

She takes another capsule of Meprospan-400 with her evening meal. She has enjoyed sustained tranquilization all day — and has had no between-dose letdowns. Now she can enjoy sustained tranquilization all through the night.

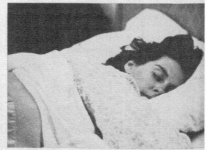

Relaxed, alert, attentive . . . she is able to listen carefully to P.T.A. proposals. For Meprospan does not affect either her mental or her physical efficiency.

Peacefully asleep . . . she rests, undisturbed by nervousness or tension. (Literature on Meprospan is available from Wallace Laboratories, Cranbury, N. J.)

Tranquilizers to help housewives do housewifely things. *Source:* Meprospan ad, *JAMA,* July 16, 1960. Courtesy of the New York Academy of Medicine Library.

Regained: the ability to make decisions, the emotional energy to complete his work

DEXAMYL®
brand of dextroamphetamine
sulfate and amobarbital

helps relieve symptoms
of mild depression
within the hour

Men and women returned to "traditional" gender roles by Dexamyl.
Source: Dexamyl ads, *AJP,* August 1965 and January 1964. Courtesy of the New York Academy of Medicine Library.

Regained: *an interest in her surroundings, a feeling of well-being*

DEXAMYL® brand of dextro-amphetamine sulfate and amobarbital

helps relieve symptoms of mild depression within the hour

Each Tablet contains 5 mg. of Dexedrine® (brand of dextroamphetamine sulfate) and ½ gr. of amobarbital, derivative of barbituric acid [Warning, may be habit forming]. Each Spansule® sustained release capsule No. 1 contains 10 mg. of Dexedrine (brand of dextroamphetamine sulfate) and 1 gr. of amobarbital [Warning, may be habit forming]. Each 'Spansule' capsule No. 2 contains 15 mg. of Dexedrine (brand of dextroamphetamine sulfate) and 1½ gr. of amobarbital [Warning, may be habit forming]. The active ingredients of the 'Spansule' capsule are so prepared that a therapeutic dose is released promptly and the remaining medication, released gradually and without interruption, sustains the effect for 10 to 12 hours.

INDICATIONS: (1) For mood elevation in depressive states; (2) for control of appetite in overweight.

USUAL DOSAGES: One 'Dexamyl' Tablet 2 or 3 times daily—in depressive states, at mealtimes; in overweight, 30 to 60 minutes before meals. One 'Dexamyl' *Spansule* capsule taken in the morning for 10- to 12-hour effect.

SIDE EFFECTS: Insomnia, excitability and increased motor activity are infrequent and ordinarily mild.

CAUTIONS: Use with caution in patients hypersensitive to sympathomimetics or barbiturates and in coronary or cardiovascular disease or severe hypertension. Excessive use of the amphetamines by unstable individuals may result in a psychological dependence; in these rare instances withdrawal of medication is recommended. It is generally recognized that in pregnant patients all medications should be used cautiously, especially in the first trimester.

SUPPLIED: Tablets, in bottles of 100; 'Spansule' capsules No. 1 (1 dot on capsule) and No. 2 (2 dots on capsule), in bottles of 50.

Prescribing information Jan. 1963

 Smith Kline & French Laboratories

VI

Antidepressants to help get her in the mood for domestic duties. *Source:* Vivactil ad, *JAMA*, April 13, 1970.

notion that America's middle-class housewives, like their white-collar husbands, were riddled with anxiety and neuroses. Friedan agreed: American women, she wrote, were haunted by a "problem that has no name," a problem whose symptomatology—the "housewife's syndrome" —might have been lifted straight out of a tranquilizer ad: "mild, undiagnosable symptoms . . . malaise, nervousness, and fatigue." Sufferers complained of feeling "empty somehow," or having "a tired feeling" or "a strange feeling of desperation." Their voices tended to have a "dull and flat, or nervous and jittery" quality.[65] These were Friedan's words, but Miltown's marketers could not have put it better.

Friedan also accepted the neurasthenic logic that linked these women's anxiety to their elevated class status. The "problem that has no name" was quite openly an ailment of affluence, its miseries not (as she took care to point out) "caused by lack of material advantages." To the contrary: its victims were women "to be envied for their homes, husbands, children, and for their personal gifts of mind and spirit." These "uncommonly smart and capable" women were well-educated but had abandoned their schooling "below the level of their ability" in order to marry and raise a family.[66] It was this unique set of circumstances—the failure to meet great mental and economic potential—that produced the vague, ambiguous anxieties afflicting them.

By suggesting that women's anxiety came from stymied dreams rather than too much ambition, however, Friedan parted ways with drug advertisers and gender traditionalists. Borrowing a page from Miltown's vociferous masculine critics, Friedan argued that women's psychic suffering was not a medical phenomenon to be soothed by tranquilizers but a political problem to be cured through social change. Women were not sick because they had abandoned the home and domesticity; rather, they were sick because they were being restricted to unfulfilling family life. In fact, one of their biggest problems was that they had listened to "experts" telling them that their nameless problems were medical illnesses. The result? They visited a physician and "blotted out the feeling with a tranquilizer." Friedan reported with alarm that "many suburban housewives were taking tranquilizers like cough drops." As one of the women she interviewed explained, "You wake up in the morning, and you feel as if there's no point in going on another day like this. So you take a tranquilizer because it makes you not care so much that it's pointless."[67]

In Friedan's eyes, such treatments were no better for women than they were for men. Like the Miltown-swallowing men who lost ambition at work or passion in politics, tranquilized women were merely dulling themselves into accepting circumstances that needed changing. The root issue, Friedan argued, was self-fulfillment. Women, like all humans, needed lives that "permitted the realization of one's entire being." A suffering housewife's "anxiety can be soothed by therapy, or tranquilized by pills or evaded temporarily by busywork. But her unease, her desperation, is nonetheless a warning that her human existence is in danger . . . If she is barred from realizing her true nature, she will be sick."[68]

Tranquilizers were far from central to Friedan's work, but it is notable that in self-consciously crafting her appeal to middle-class women, she found the pills a convenient shorthand for the social—and medical—forces that kept housewives in their place. The dozen or so times she addresses tranquilizers helped attach her arguments to the neurasthenic logic being promoted by powerful forces such as the marketing engines of commercial medicine. Most of her points had in fact already been made, repeatedly, and would have been familiar to anyone who had read a *Cosmopolitan* article about the new wonder drugs, seen a drug advertisement, or possibly even talked to their own doctor about anxiety. Friedan's task was to take this familiar and well-circulated material and rearrange it to support a very different conclusion. In embracing the neurasthenic tradition in this way, of course, she restricted herself to the upper rungs of America's economic ladder—but this had been her goal from the start. Ironically, this towering early figure in second-wave feminism cast herself as the partner, not the critic, of Cold War masculinists. Like them, she argued for the existence of an essential middle-class self at risk in modern America, one that required political care, not drugs, to achieve authenticity and fulfillment.

The Feminine Mystique reverberated powerfully enough through American culture that by the end of the 1960s, at least some tranquilizer advertisers appeared to respond directly to Friedan's critique. In 1968, for example, Miltown's advertisers ran a campaign focusing on "syndromes of the 1960s." The first of these ads, the "battered parent syndrome," depicted a tiny woman terrorized by a huge child wielding giant-sized alphabet blocks (they spell "tension"). According to the accompanying text, tranquilizers could help ease the Friedanian anxiety of educated women plagued by their mundane domestic existence. "Some say it's unrealistic to educate a woman and then expect her to be content with the Cub Scouts as an intellectual outlet," ran the copy. Miltown might be "no substitute for a week in Bermuda, or for emotional readjustment. But it will often make the latter easier for her, as well as for the physician." A 1969 ad for the minor tranquilizer Serax commiserated with a woman pictured as trapped behind bars made of brooms and mops: "You can't set her free. But you can help her feel less anxious."[69]

On one level, this kind of advertisement simply continued the gendered logic pervading early tranquilizer ads: if women were unhappy in the home, they should "readjust" to the situation, possibly through the use of tranquilizing drugs. As commercial marketing tends to do, these ads identified deviance as a problem to be solved by adapting to the environment rather than by changing it. One would not expect—nor would one find—an ad in which a downcast businessman learned to do

Solving "the problem that has no name" with tranquilizers. *Source:* Miltown ad, *JAMA*, January 22, 1968.

housework, a housewife discovered her powerful competitive spirit, or any other inappropriate behavior, gendered or otherwise. Women's unhappiness was a problem to be solved at the individual rather than the social level, by the purchase of a product. A demeaning work (or home) environment, after all, could not be "cured" through tranquilizers.

If the advertisements' overt message was that housewifely roles ought to be fulfilling, however, they also—perhaps unintentionally—spread and expanded Friedan's logic. Ostensibly, they affirmed domestic wifeliness as normal and healthy. At the same time, however, they argued that women needed pills to achieve this "normal" state. By presenting universal images of generic-looking housewives in classic domestic settings, they also implied that most if not all women were discontented and had to be "helped" by drugs to achieve the desired normality. This was in keeping with the basic goal of selling more drugs, but it also promoted the idea of housewives' dissatisfaction. Madison Avenue's powerful visual and rhetorical strategies mapped the vocabulary of illness onto the geography of women's everyday life even more acutely than Friedan ever did. Women were portrayed as unhappy not in some abstract, clinical, or biological way but almost always in archetypal domestic settings. The intent was to illustrate women's pathologies, but the ads could also be seen as figuratively joining them in being dissatisfied with the countless burdens and constraints of enforced housewifery.

Ironically, then, advertisers' universal portrait of middle-class American women as vaguely discontented and in need of psychotropic help broadened an already opening space for an alternative interpretation of "women's complaints." Just as drug critics helped circulate advertisers' claims, as we saw in chapter 1, advertisers in turn helped circulate critics' charges. The notion that tranquilizers and antidepressants were women's drugs served a variety of agendas, from selling drugs, to supporting the normative housewifely role, to opposing that same role.

Together, medical practices, advertising, and cultural criticism helped make Librium and Valium the quintessential women's drugs of the 1960s and 1970s, respectively. At the same time, however, advertisers continued to seek male customers by helping to amplify and circulate biological notions of the primeval and powerful masculinity lurking in even the most apparently overcivilized of men. These cultural processes played a crucial role in writing the 1950s gender crises into the emerging sciences of biological psychiatry and sociobiology.

Eventually this dynamic back-and-forth would open new opportunities for feminists, especially those in the women's health movement, and Friedan's argument would become much more widespread. But not yet. Part of the reason Miltown, Librium, and Valium served so well in these early debates was their presumed effectiveness: even during this supposedly Freudian era, critics took for granted that the drugs worked, that they genuinely could control minds. By the later 1970s this was no longer the case. Fears of mind control had shifted to fears of minds out of control, as a panic developed over a supposed national epidemic of Valium addiction among white middle-class housewives. To understand how and why this shift occurred, we must look at a set of debates about tranquilizers and drug dependency that had been developing quietly, in the relative obscurity of limited medical circles, from almost the first days of Miltown's existence.

CHAPTER THREE

Wonder Drugs and Drug Wars

FREDERICK LEMERE, a psychiatrist at the University of Washington, was worried. He had been an early enthusiast of Miltown, prescribing it for over six hundred patients in 1955. His glowing report on the medication —one of the earliest to be published—concluded that it was "the drug of choice for the relief of tension, anxiety, and insomnia."[1] By 1956, however, he had noticed a disturbing phenomenon: a "small but definite" proportion of his patients were taking it upon themselves to increase their dosage, often to excessive levels. Their families complained that they seemed drunk, and Lemere himself observed patients on increased doses manifesting "all the signs of intoxication," including euphoria and poor coordination. When he took these self-medicating patients off Miltown, they complained of "a feeling of 'nervousness' and 'the jitters,'" and one patient who had reached a very large dosage actually suffered a convulsion.

For a psychiatrist whose practice exposed him to alcoholics and other drug abusers, the signs seemed easy to read: meprobamate could cause addiction. With prescriptions for the drug already in the tens of millions per year and increasing rapidly, this was a troubling discovery, and Lemere dutifully reported his findings to both the *Journal of the American Medical Association* and the *AMA Archives of Neurology and Psychiatry*. "None of this is intended to depreciate the clinical value of meprobamate," he concluded carefully, "but it does behoove the physician to ponder the implications of a mass dependency on this drug."[2]

As cautious as Lemere had been in his choice of words, his warning would still have sounded alarmist in many quarters in the 1950s. And even those inclined to accept it would have been hard pressed to say exactly what could or should be done about it. Tranquilizers were medicines, after all, not dope. And for over three decades an antidrug coalition of experts and police authorities had enshrined clear lines be-

83

tween the two: "dope" meant illegal, addictive drugs sold by "pushers" to nonwhite, poor, or otherwise marginal populations; "medicines" were legal life-savers sold proudly to physicians by "merchants of health." Because drug policy had emerged in tandem with efforts to patrol urban vice, these distinctions emphasized types of drug users rather than types of drugs. Few antidrug crusaders had reason to question them, or to think of "addict" and "patient" as anything other than mutually exclusive categories.[3]

But new developments in the 1950s had begun to place this bifurcated system under increasing stress. First, Lemere was not the only—or even the first—voice to raise suspicions about prescription medicines. Many of the nation's most influential drug researchers had been reporting dependence on other sedatives and stimulants for over a decade. Second, the relentless commercialization of medicine led to ever-increasing sales of the medicines under scrutiny, raising the possibility of "mass dependency" virtually inconceivable in the prewar era. And finally, the political climate surrounding addiction had begun to change by the late 1950s. Alcoholics Anonymous and a new generation of medical specialists advocated understanding and treatment for addicts, undermining the assumption that drug use was a criminal act requiring policing and punishment.

Together these dynamics began to fracture the decades-old antidrug coalition, forcing a reexamination of difficult questions about addiction. Did addiction result from exposure, or from a criminal predisposition—in other words, was addiction a disease or a crime? Or, to put it another way, were drugs addictive, or were people disposed to addiction? Was all addiction necessarily a threat to society, or only when it occurred among dangerous people? Should addiction be measured by pharmacology, or by social harm? Each of these troublesome questions had been settled, if arbitrarily, by antidrug experts and campaigners as far back as the 1920s, and each now came under new scrutiny as these same experts and campaigners began to reevaluate the hard and fast lines between medicines and dope.

The reemergence of these questions was not the result of radicals seeking to question or dramatically alter America's antidrug policies. Rather, they represented conservative efforts by those within the antidrug coalition to extend controls over prescription drugs. They sought to add a new category of "addiction" to America's drug lexicon without disturbing the existing one. Indeed, it was the concerted attempt to make changes without undermining existing policy that made the task so difficult for the drug researchers, addiction specialists, and political campaigners who undertook it. No one wished to suggest that a respectable patient addicted to a sedative was somehow morally, socially, or politically equivalent to someone injecting heroin "on the street"—or, more to the point, that the two should be legally equivalent. Their goal was to create new safeguards for prescription medicines without repeating the process that had outlawed a previous generation of medicines (opiates and cocaine in particular).

Making this agenda even more difficult to achieve was the increased political and social power of professional medicine and the pharmaceutical industry. Both of these groups claimed the mantle of "expert" authority over drugs and resisted any effort to cede that authority. They had at their command the far-reaching networks of commercialized medicine, which provided a substantial infrastructure for producing and circulating favorable information about prescription medicines. If there was any ambiguity about the addiction risks of these medicines—and there was— it would be aired, thoroughly, by these powerful defenders.

Lemere's warning of "mass dependency" on Miltown represented a significant challenge in this context. He and other reformist experts who criticized tranquilizers were making a new kind of argument: all drugs that sedated (or stimulated) carried the potential for dependence and abuse, regardless of who used them. Addiction hinged on pharmacology, and pharmacology was no respecter of laws or social categories. This reasoning was not entirely new, but in applying it to medicines with little reputation for illegal use, reformers had certainly upped the pharmacological ante. Again, this did not mean upending existing drug policy. As reformers hastened to assure, just because all potentially addictive drugs needed some sort of control did not mean that they all needed to be controlled in the same way. Nevertheless, these were significant changes that opened a new and important phase in a battle just begun over the very meaning of addiction—the first major reconsideration of the concept since the origins of drug control in the 1910s. The stakes were high.

Ultimately, tranquilizers like Valium were indeed brought under federal police authority using such pharmacological language, but the victory was neither swift nor clean. Medical, pharmaceutical, and government authorities sparred interminably over the meaning of addiction in the halls of Congress, on the pages of medical journals, and through the complex networks of commercial medicine. The debates raged for nearly twenty years, ending only in 1975 when Valium, the last holdout, was placed alongside narcotics and other kinds of dope on the newly formed Drug Enforcement Administration's "Schedule of Controlled Substances." Thanks to intense medical and trade lobbying—and skillfully exploited ambiguities in the concept of addiction—the tranquilizer occupied a favored spot, Schedule IV, created especially to separate it from illegal narcotics.

Beyond helping us understand the changing reputations of the one-time wonder drugs Miltown, Librium, and Valium, this saga provides an important glimpse into the contested historical processes that have created and managed both expert and commonsense knowledge about addiction and American drug policy. Miltown, Librium, and Valium strained the explanatory power of existing cultural frameworks about addiction. The complex effort to manage this challenge intensified differences among antidrug experts and revealed the unspoken ways in which the medical logic of addiction had become intertwined with a policing agenda.

This chapter begins with a brief look at the origins of distinctions between "drugs" and "medicines," then goes on to examine how those distinctions were challenged, redrawn, but at the same time reinforced during the decades-long process of controlling Miltown and Valium. This was a long, slow drama, met at each step by conflict and uncertainty. It began with contentious congressional hearings on prescription sedatives (barbiturates) in the 1950s, continued through experts' redefinitions of "addiction" in the early 1960s, and concluded with a decade-long struggle over whether to include Valium under a new, pharmacology-based federal drug control regime inaugurated in 1965. At each stage, physicians, drug companies, addiction experts, and antidrug crusaders struggled to define medical truth about drug abuse without contradicting the political and cultural imperatives that also drove American drug policy.

Before telling the story, it may be useful to clarify my own terminology, especially since this chapter argues that the meanings of key terms such as "addiction" have changed over time. In this chapter, *physical dependence* refers to a physiological habituation to a drug, such that stopping use will cause unpleasant physical and psychological symptoms known as *withdrawal* (or *abstinence syndrome*). Typically accompanying this dependence is the phenomenon of *tolerance*, referring to the need to take increasing amounts of a drug to achieve the same effects. Finally, *addiction* itself will refer to medical, legal, or also popular definitions of drug dependence; its meaning thus changes depending on the historical context in which it was used.

The Street and the Medicine Cabinet

Before turning to the effort to regulate minor tranquilizers in the 1960s and 1970s, we first need to understand the process by which street drugs like heroin, cocaine, and marijuana came to be medically and legally distinguished from prescription drugs. As this history will show, there are no clear and consistent pharmacological explanations for historical drug categories; instead, drug policy emerged out of conflict and cooperation between physicians, Progressive-era reformers, and working-class men in urban vice districts. The result was a set of medical concepts of drugs, medicines, and addiction thickly interpenetrated with political ideologies about the "dangerous classes" and alien threats to native-born, middle-class, respectable American culture.

During the nineteenth century, the "dope menace" took a back seat to "Demon Rum." Temperance was the main antidrug crusade, its thousands of partisans battling the evils of alcohol and, especially later in the century, the working-class ethnics presumed to be under its sway. Morphine, meanwhile, was a godsend dispensed legally by physicians to respectable patients or sold directly to consumers as patent medicines. True, some observers were already denouncing opium-smoking "Orientals" in the West and, after the Civil War, drug-crazed African-American men supposedly running rampant in the South. But no organized antinarcotic campaign emerged to match Temperance, and for the most part, medical therapeutics remained the primary frame for understanding opiates and other drugs. Addiction was considered a wretched or even contemptible state, but it was understood to be a private affair of the middle and upper classes—hardly a cause for widespread alarm.[4]

Divisions between "medicine" and "dope" began to emerge around the turn of the century, as the dangers of addiction became better known and the demographics of drug use shifted. The American Medical Association (AMA) campaigned vigorously to reduce physicians' prescriptions of opiates, and muckraking journalists exposed the worst practices of patent medicine companies. By the early twentieth century, these campaigns had scored at least two significant victories: the establishment of the Food and Drug Administration to enforce truth in labeling, and the Harrison Anti-Narcotic Act, which initiated federal control over narcotics. These victories dried up the two main legal sources of drugs (doctors and patent medicines), and the number of addicts from the respectable classes dropped dramatically. This left other, previously less noticeable groups to become the public face of American drug use: immigrants, urban workers, petty criminals, and others in the racially mixed milieus of America's burgeoning urban vice districts. As these vice districts came under increased police vigilance during the Progressive era, addicts maintaining a habit were forced ever deeper into a shady criminal economy—a development that only seemed to further prove their deviance and immorality.[5]

By the 1920s, drugs, now including heroin and cocaine, had become fully associated with urban "dangerous classes." At the same time, the ongoing antidrug campaign changed in character from a medical to a police affair. No longer was it an internal AMA effort to reduce prescriptions or a muckraking crusade against corrupt patent medicine companies. Instead, new laws empowered police to assume responsibility for drug control. To lead the new antidrug effort, Congress established the Federal Bureau of Narcotics in 1929, headed by the publicity-savvy Harry J. Anslinger. The Supreme Court agreed that these federal narcotics agents had the power to determine whether a physician who prescribed an opiate did so "in good faith" (i.e., legally, to cure an illness) or simply to maintain an addict (i.e., illegally). Police, not physicians, would decide what a "real" prescription was.[6]

Addiction researchers were central to the new antidrug regime. Studying drug users' criminalized world of vice and deviance in the 1910s and 1920s, they came to be increasingly pessimistic about the prospects for treatment. Their gloomy assessments pushed medical wisdom away from seeing addiction as a disease—something that anyone could contract if exposed to it—and toward seeing it as a consequence of addicts' preexisting character flaws. Their influential work described deviance, crime, and vice as elements of addiction itself, not strategies for maintaining a habit already defined as criminal. The nation's most prominent drug expert, psychiatrist Lawrence Kolb, argued that addicts were psychopaths long before they had ever used a drug. Indeed, he went so far as to deny that narcotics gave "stable persons" any pleasure whatsoever. Despite—or perhaps because of—recent professional memory of "medical addicts," Kolb and other physicians abandoned addicts as patients, instead producing medical knowledge that supported the Narcotics Bureau's demonology of violent and deranged "junkies."[7]

As police assumed responsibility for drug policy in the 1920s, they intensified the shift away from understanding drugs to controlling drug users. Narcotics Bureau chief Harry Anslinger led an enormously influential public-relations campaign blaming desperate and violent dope fiends for robberies, muggings, and urban moral degeneracy. Similar campaigns unfolded on local stages as well. In Los Angeles, for example, the police department and local media worked together to sensationalize the drug threat as a way to popularize the need for tight police control over the city's activist Mexican workers. News programs and movies helped the police department circulate stories of heroic officers battling demented marijuana-addicted Mexican men. In these and other cases, antidrug campaigns became vehicles for two seemingly complementary agendas: limiting addiction and policing the "dangerous classes." One result was the association of addiction with nonwhite, poor, or otherwise threatening social groups—not just in popular culture, but in the interlocking networks of police and medical authorities who produced expert knowledge about drugs. The antidrug coalition of police, cultural crusaders, and physicians helped make it seem a logical absurdity for an ordinary, respectable (i.e., white, middle-class) American to become addicted unless they had been "infected" by someone from the dangerous classes.[8]

Lost in this campaign were meaningful distinctions between different types of drugs. All illegal drugs (dope) were considered addictive. Marijuana, for example, did not foster physical dependence in the manner of heroin, but it was nonetheless included as one of the "habit-forming narcotic drugs" whose users could be consigned to a federal prison/

hospital for drug offenders, and "marijuana addicts" featured prominently in Hollywood drug films and other antidrug campaigns.[9]

Prescription drug addicts, meanwhile, whose respectable class status and nonthreatening aura might have complicated the moral clarity of the Narcotics Bureau's task, were essentially ignored by the antidrug coalition. Physicians, well rid of the stigma of iatrogenic addiction, saw little reason to challenge this approach. And pharmaceutical companies, of course, vigorously defended their products, just as they had (unsuccessfully) defended opiated patent medicines earlier in the century. By the height of the narcotics scares in the 1950s, "addictive" drugs had come to mean—both legally and culturally—those used by "addictive" populations.[10]

The Street Meets the Medicine Cabinet: The Case of the Barbiturates

What the new antidrug consensus largely failed to acknowledge was that the old nineteenth-century drug culture—respectable white women and men receiving sedatives and energizers through physicians' prescriptions —had not disappeared with the regulation of narcotics. Instead, it had simply moved on to newer and less stigmatized drugs: the barbiturates, a class of sedatives introduced in 1903; the amphetamines, a class of stimulants introduced in 1932; and, eventually, the tranquilizers and antidepressants. Protected by the positive aura of medical therapeutics and powered by the increasingly commercialized drug industry, these medicines were already among the most popular drugs in the pharmacopeia well before their spectacular postwar rise.[11]

The seemingly limitless climb in use of these drugs in the 1940s and early 1950s raised unexpected questions about American drug policy from within the antidrug consensus. Drug researchers and some antidrug crusaders, concerned about the potential addictiveness of prescription sedatives and stimulants, began a slow and often tentative push to understand addiction as a problem of pharmacology as well as character flaws, something that certain kinds of substances could produce in anyone, not just psychopaths. Their effort culminated in a push to regulate the barbiturates when Congress tightened the nation's drug laws in 1955.

The push did not succeed, but it did lay the groundwork for more successful challenges to the medicine-drug divide in the 1960s.

Before this midcentury push, regulation of prescription sedatives and stimulants had followed a different course from the battle against dope. Free of the sensationalism of antinarcotics discourse, they did not come under any sort of federal control until 1938, and even then regulations were nominal. The Food and Drug Administration declared them to be "habit forming," finding that they could produce psychic but not physical dependence (and thus not addiction): people came to like, but not need, the "pleasurable stimulation or escape from unpleasant experience" they offered. It was not until 1951 that FDA policy was formalized by new legislation restricting sedatives and stimulants to a physician's prescription. The new law also required a physician's written consent to refill a prescription, and it mandated a warning label ("may be habit forming") for medicine bottles. The law did not require manufacturers or retail sellers to keep records, however, and it had other limitations too: it set no limits on personal possession, established no special protections for juveniles, and applied only to trade between—not within—states. Like the loose and uneven patchwork of state laws, moreover, the new federal regulations were respectful of medical authorities, in stark contrast to antinarcotic laws that permitted government authorities to judge whether or not a prescription served a legitimate therapeutic purpose. This selective willingness to defer to medical authorities helped foster an enduring split between medicines and dope.[12]

By midcentury, however, links between prescription medicines and addiction were increasingly (if carefully) being raised from sources not easily ignored: the same addiction researchers whose "proof" of addicts' psychopathology had been a pillar of the antidrug consensus. In addition to producing gloomy scholarship on addicts' innate character flaws, these researchers had also been tasked with discovering a nonaddicting painkiller. If they succeeded, authorities could abandon the hopeless quest to cure addicts; instead, they could control the drug problem by ending the supply of opiates altogether. Because this goal assumed that addicts were a lost cause, it harmonized well with the punitive policies of the antidrug crusade.[13]

At the same time, however, the search for a nonaddictive painkiller led researchers to explore the addictiveness of prescription medicines in ways that ultimately proved destabilizing to the Anslinger-era consensus. Nathan B. Eddy's research at the University of Michigan, for example, pursued under the auspices of the National Research Council's Committee on Drug Addiction, ultimately convinced him—against his own initial hopes—that every new supposedly nonaddictive morphine-type medicine released with much fanfare by pharmaceutical companies produced physical dependence just like morphine.[14] Colleague Harris Isbell continued similar research for the Public Health Service's Addiction Research Center at the federal narcotics "farm" in Lexington, Kentucky (founded in 1935, the Lexington hospital/prison for addicts included a working dairy farm as part of inmates' therapy, and it also housed the nation's premier research unit on addiction). As one Public Health Service official explained in 1955, "Drugs are constantly being developed by drug houses and tested at Lexington, a dozen or more a year. These drugs are sent down to Dr. Isbell to test to try to find out whether they are more or less addicting . . . so far no one has ever obtained a drug which would have the same pain-relieving effects as morphine without the addiction liability."[15]

Pushing against major drug companies in this way helped Eddy, Isbell, and other researchers acknowledge the addictiveness of prescription sedatives and tranquilizers, even in the face of corporate resistance and even when their findings contradicted the simple dichotomies of U.S. drug policy. It was Isbell, for example, whose experiments in the 1940s proved that barbiturates caused physical dependence.[16] His colleague Carl Essig, meanwhile, confirmed Frederick Lemere's troubling observations about Miltown, reporting in *JAMA* in 1957 that he had witnessed "hyperirritability of the central nervous system and convulsions" when patients stopped taking high doses of the drug. Even more damning, he had seen similar symptoms in dogs withdrawn from the drug, three of whom had suffered fatal seizures.[17] The same year, Eddy published a more general warning in *JAMA*: if taken in adequately large amounts, all drugs, old or new, offering "some easement of tension . . . tend to promote a habit in those who take them" and could "set into operation the phenomena of physical dependence."[18] The United Nations' Commission on Narcotics Drugs weighed in too, noting with concern the "rapidly-increasing" use of "tranquillizing" drugs and officially resolving that they should be "classed as potentially habit forming" and "subjected to national control."[19]

These were hardly marginal voices. Isbell directed research at the nation's most important facility for addiction studies at the Lexington narcotics farm, where Essig also worked. By the 1950s Eddy had become an enormously influential figure as chief of the Section on Analgesics (painkillers) at the National Institute of Health and chair of the World Health Organization's Expert Committee on Addiction-Producing Drugs. Theirs were serious warnings from central institutions of the antidrug coalition, and they did not go unheeded.

The first major sign of success came in 1955, when Congress held extensive hearings on the barbiturates. These hearings are known primarily for producing the harshest federal antinarcotic law in the nation's history, known as the "little Boggs bill" because it strengthened an already harsh 1951 law championed by Louisiana's crusading antidrug congressman Hale Boggs. The new Boggs bill added stiffer mandatory jail time and the death penalty for pushers who sold to juveniles. It represented the height of the Anslinger-era consensus, a time when attitudes toward addicts had been hardened even further by the Cold War animus against all forms of social deviance.[20] But the proceedings were officially called *Traffic in, and Control of, Narcotics, Barbiturates, and Amphetamines*, and more than half of the thousands of pages of transcripts relate directly to prescription sedatives and stimulants. The intentions of the committee, chaired by Boggs, were clear: publicly air growing fears about barbiturates (the most popular sedatives) in the hopes of gathering the political will to regulate them more tightly. These hearings revealed the growing strains in the antidrug coalition, but they also showed the continuing power of the logic dividing medicines from dope.

Divisions among antidrug experts were most evident in the august roll of influential figures testifying against barbiturates. Assistant Surgeon General G. Halsey Hunt and Kenneth Chapman, consultant on addiction for the Public Health Service's National Institute of Mental Health, set the tone with opening testimony that addiction to barbiturates was "more serious than narcotics addiction" and "more difficult to treat." Asked if marijuana was any worse than barbiturates, Chapman returned an unambiguous "No sir." Senators were shown an instructional movie from the Lexington narcotics farm that included graphic and deeply disturbing images of patients thrown into violent psychosis during withdrawal from barbiturates. Harris Isbell provided a copy of all his barbiturate research and described in grim detail the possibility and consequences of addiction. Adding a sensational edge to this testimony were ominous reports of what seemed to be massive overproduction of the medicines, beyond even their rapidly growing legitimate use. Jerome Trichter, assistant commissioner of environmental sanitation of New York City's Department of Health, reported that the ranks of barbiturate addicts were exploding, with fatal poisonings going up by a factor of ten between 1938 and 1954—far outpacing morphine.[21]

Despite these dire warnings, however, Congress did not tighten control over barbiturates and amphetamines, even as the Little Boggs Act added the death penalty for narcotics offenders in 1956. The most obvious reason was the determined opposition of commercial and professional interests, including virtually every organized medical, drug industry, and druggist trade group: the American Medical Association, the American Pharmaceutical Association, the American Pharmaceutical Manufacturers Association, the National Wholesale Druggists' Association, and even the American Dental Association.[22] One FDA investigator later testified directly about their impact, telling the Congress in 1961 that Albert Holland, the FDA's medical director, had ordered her to "do nothing about" new information regarding Miltown's addictiveness in 1957 because he did not want his "policy of friendliness with industry interfered with."[23]

Why were these commercial and professional concerns able to resist federal control over potentially addictive drugs at the height of America's antidrug fervor? Part of it can be explained by the increasing strength of medicine and the pharmaceutical industry in a golden era of medical advances and commercialization. They simply had more political influence and respected expertise to bring to the table. At the same time, however, they were successful because they were able to exploit ambiguities in the meanings of addiction. The term had grown to describe what happened to junkies—marginal men, criminals skulking in the demimonde of urban vice districts. Whatever dangers might be posed by barbiturates and other prescription medicines, they were not difficult to portray as fundamentally different.

A prime example of this ambiguity is Harris Isbell himself, the researcher centrally responsible for proving that barbiturates could indeed be addictive. As we saw, Isbell obligingly provided his experimental conclusions and a raft of publications on the dangers of barbiturates. But senators hoping he would come down strongly in favor of new regulations were disappointed. "The fact that people abuse these drugs," he testified, "does not really mean there is anything wrong with the drugs . . . it really means that there is something wrong with the people who use the drugs improperly." The "personality disorder" that caused addiction, moreover, almost always led people to choose other drugs over barbiturates: "All the barbiturate addicts that I have seen have been secondary to either opiate addiction or alcoholism. I have never seen a patient in whom the barbiturate was the first drug of addiction." Ultimately, Isbell argued, "in spite of the fact that I am responsible, personally responsible, for showing that, with respect to physical damage to the individual and to society, barbiturate addiction is worse than morphine addiction, I am not in favor of putting barbiturates under the narcotics laws." Barbiturates, he argued, were medically useful, relatively safe in that there were few addicts despite wide use, and—what's more—would be immediately replaced with other, possibly worse, sedatives even if they were outlawed.[24]

Isbell was not alone among drug experts in resisting parallels between narcotics and medicines. Kenneth Chapman, the addiction specialist at the National Institute of Mental Health whose opening testimony had so powerfully painted the dangers of barbiturate addiction, backed down when it came to the question of federal regulation: "I am not so sure whether it is a matter of regulation. Perhaps it is a matter of education as to the effects . . . and prescribing quantities."[25] AMA witnesses agreed: Maurice Seevers of the AMA's Council on Pharmacy and Chemistry testified that "we would not like to create another Volstead Act situation [i.e., Prohibition] with the barbiturates, nor would we wish to put barbiturates into the same category as those under the Harrison Narcotic Act," because, despite the increasing instances of addiction to barbiturates, "they in many instances apply only to noninjurious habits . . . [more] comparable to alcohol than morphine."[26] Drug industry representatives also proposed a "model bill" that kept red tape for legitimate providers to a minimum while focusing almost exclusively on tracking down illicit trade. Once again this was a very different matter from antidrug laws, which gave police the power to define a legitimate prescription.[27]

Anslinger himself personally resisted extending his bureau's jurisdiction to cover prescription drugs, telling the committee, "I think this can be worked out by the medical profession and the pharmacy profession also. It is on their doorsteps now." The problem, he explained, was a lack of flexibility once the bureau was given jurisdiction over a drug: "Once we take hold of it, all of our narcotics laws put everything in the same compartment as morphine . . . if you want to control a drug you have to really control it." Such enforcement would have unfortunate consequences for widely used medicines like the barbiturates, in that it would bring ordinary, respectable Americans into the criminal justice system. Such tactics, Anslinger believed, had doomed the nation's experiment with Prohibition and would undermine popular support for drug control, too.[28]

Anslinger's attitude recapitulated the broader contours of the antinarcotic campaign, the effectiveness of which relied on moral simplicity and social prejudices—both of which might be undermined by acknowledging respectable medical addicts. These simple dichotomies made it difficult to acknowledge and grapple with addictiveness to prescription medicines like the barbiturates, even for researchers like Harris Isbell. Making matters even more complicated were professional and trade groups, which played up the ambiguity of addiction to defend widely used medicines. Since people who became dependent on barbiturates did not seem to act like junkies, they did not merit control with the punitive measures imposed on narcotics earlier in the century. Instead, education and stricter adherence to physicians' authority were the preferred solution. Barbiturates had brought out differences within the antidrug coalition, but not enough to break the basic agreement about lines between the street and the medicine cabinet.

Rehabilitating Addiction and Legislating Pharmacology

A decade after failing to insert barbiturates into the little Boggs Act, antidrug reformers finally notched a legislative victory with the passage of the Drug Abuse Control Amendments. The 1965 law imposed new federal controls over barbiturates and amphetamines and explicitly called for investigation and control over tranquilizers like Miltown and Valium. Its language clearly identified pharmacology as the basis for the new regulations: the substances it covered were those with a "potential for abuse" because of their "sedative or stimulant effects." In acknowledging that abuse might follow exposure to legal medicines, this took a significant step away from the junkie paradigm and toward a disease model of addiction. As we shall see, however, this step toward legislating pharmacology was highly contested and partial. It directly regulated barbiturates and amphetamines, which had been increasingly identified with illegal use by suspect populations, while leaving the more respectable tranquilizers (Miltown and Valium) for further deliberation.

To understand the new law and its limits, we must first examine the political and medical changes in the late 1950s and early 1960s that made it possible. One of these was the rise of Alcoholics Anonymous, which, along with the reemergence of physicians determined to treat rather than punish addicts, breathed new life into the long-defunct disease model of addiction. Meanwhile, drug researchers began to put forward a more confident pharmacological explanation for how drugs caused addiction. By tending to emphasize pharmacology over character flaws, these developments "rehabilitated" addiction into something that might happen to anyone.

Perhaps the best known of these developments was the emergence of the disease model of addiction in the 1950s, as a new generation of addiction treatment specialists questioned the assumption that all addicts were depraved and vicious junkies. Appalled by the punitive policies at Lexington, pioneers like Dr. Marie Nyswander claimed that addicts were sick, not evil, and could conceivably be treated and even cured by physicians' "intelligent and understanding help." Nyswander was one of the first to advocate providing heroin addicts with the long-acting opiate methadone so that they could lead relatively normal lives free of the need to maintain an illegal habit. Given the political context, this was an optimistic approach that assumed addicts wanted to, and could, lead such lives if freed from the necessity of constantly purchasing illegal narcotics.[29]

Individual physicians were not the only ones questioning American antidrug policy. In 1961 the American Medical Association and the American Bar Association publicly challenged the harshly punitive approach of the past decades, insisting that physicians and judges should not be prohibited from allowing treatment for addicts.[30] Harry J. Anslinger, the brilliant politicker and the human dynamo at the heart of so many antidrug campaigns, retired in 1962. That same year, in *Robinson v. California*, the U.S. Supreme Court declared addiction to be a disease and thus not in itself a punishable offense (although procuring, selling, possessing, and using drugs remained illegal even for addicts).[31]

Physicians' campaign to reconfigure addiction as a disease was strengthened by a new awareness that addiction struck not only marginal, stigmatized populations but ordinary middle-class whites, too. A central vehicle for this realization was the "modern alcoholism movement," an overlapping network of researchers and activists who sought to spread word that alcoholism was a disease rather than a moral state. The new "movement" included researchers at universities like Columbia and Yale, along with private organizations like the National Council on Alcoholism and the Research Council on Problems of Alcohol. Also a key institution was Alcoholics Anonymous (AA), the original self-help group for problem drinkers, which had grown from its humble 1935 origins into an organization with high visibility and tens of thousands of members. With its membership of respectable men and women, AA helped drive home the point that alcoholism was a common problem even among otherwise successful people. AA's focus on the need to rebuild the drink-damaged masculinity of male alcoholics, moreover, made the message a familiar and acceptable one during a decade obsessed with the threat of male "softness" (see chap. 2). Perhaps not coincidentally, the organization was one of the earliest to recognize and denounce the addictiveness of minor tranquilizers like Miltown.[32]

The new doctrines gained official recognition when President John F. Kennedy convened a blue-ribbon Advisory Commission on Narcotic and Drug Abuse in 1962. Its influential report debunked the notion of addicts as fearsome junkies. Addicts, the authors explained, could be "talented, even brilliant" individuals driven to drugs by fear of failure; white- or blue-collar workers seeking relief from the "tedium of their jobs and their lives"; patients accidentally addicted to medicines; workers needing drugs to "offset fatigue" on the job; and unfortunates ensnared after trying drugs "for kicks." The commission noted that at least some of these innocent addicts were not hooked on "narcotic drugs and marihuana" but "to an increasingly alarming extent" were using "other drugs such as the barbiturates, the amphetamines and even certain of the 'tranquilizers.' "[33]

The increased attention to alcohol and prescription medicines, and the acknowledgment of respectable addicts, made it easier to imagine addiction as a property of drugs rather than of people. It also opened up an unlikely connection between drug-war reformers like Nyswander and pillars of the antidrug consensus like Isbell, Eddy, and Essig. As the antidrug researchers continued to test prescription painkillers, sedatives, and tranquilizers, they moved beyond simple dichotomies of "habit forming" versus "addiction producing" to more sophisticated categories based on pharmacology. These new frameworks encompassed prescription medicines while still distinguishing them from illegal drugs, providing the scientific and cultural flexibility to address the growing divisions in antidrug thinking.

The impact of the new pharmacological research can be usefully tracked in the evolving pronouncements of the World Health Organization's (WHO) Committee on Addiction-Producing Drugs, written by or in consultation with Eddy and Isbell. The WHO's original definition of addiction, written in 1952, was simple and somewhat menacing: "(1) an overpowering compulsion to continue taking the drug and to obtain it by any means; (2) a tendency to increase the dosage; and (3) a psychic and generally a physical dependence on the effects of the drugs."[34] The first of these, with its ominous overtones—"overpowering" and "by any means" —echoed the junkie paradigm, calling to mind violent and depraved drug fiends. This language did not explicitly exclude prescription drug addiction, but the implication of criminality and social threat made it much less likely to be associated with medicines and their respectable users— as the 1955 Boggs Act hearings demonstrated.

Five years later in 1957, Eddy and Isbell revised the WHO guidelines to incorporate growing concerns about barbiturates and other prescription medicines. This involved two changes. First, they added a fourth element to the definition of addiction: the drug must produce "an effect detrimental to the individual and to society." Then they created a separate category called "habituation," whose characteristics were distinctly less frightening: "(1) a desire but not a compulsion to continue taking the drug for the sense of improved well-being which it engenders; (2) little or no tendency to increase the dosage; (3) some degree of psychic dependence on the effect of the drug, but absence of physical dependence and hence of an abstinence syndrome; (4) detrimental effects, if any, primarily on the individual."[35]

The new guidelines created a theoretical space for nonthreatening people dependent on drugs, a category easier to imagine occupied by prescription medicine users. It did not automatically carry the stigma of deviant or criminal activity; habitués were not Anslinger-era drug fiends, destroying society with their depraved "compulsion" to get narcotics "by any means." They were people who "desired" drugs upon which they had come to rely on to "some degree" because they provided a "sense of improved well-being"—that is, they treated a problem. Prescription sedatives and relaxants, Eddy reassuringly explained, posed "for most persons" a risk "only of some degree of habituation."[36] A *JAMA* summary of the new guidelines completed the logical circuit by arguing that "habituation with these [sedative or tranquilizing] agents is not primarily or essentially an abuse."[37] This framework acknowledged the danger of prescription medicines without equating them with the monochromatically evil narcotics.

Although these guidelines opened up a space for respectable, medically-induced drug dependence, they still sustained the bifurcated social logic of drug abuse. At what point does a "desire" become a "compulsion"? When does harm to an individual lead to social harm? How are such dynamics affected by the legal availability of the drug in question? The categories supposedly reflected differences between types of drugs, but they seemed better designed to capture distinctions between two different types of drug users—especially since, by Eddy's own admission, one could become "truly" addicted to even a "habit-forming" drug if one were a persistent enough user. Such instances, presumably, would be the fault of the user rather than the drug.[38]

Several years later in 1964, as researchers continued to wrestle with these and other thorny problems, the WHO fundamentally reorganized its definitions to center around drug pharmacology. New guidelines substituted the single term "drug dependence" for the overly simplistic categories of "habituation" and "addiction," and they distinguished between dependence on different types of drugs. Dependence, the WHO committee maintained, "varies with the agent involved" and should properly be described as being of a particular type, that is, "drug dependence of morphine type, of barbiturate type, of amphetamine type, etc." The distinct pharmacology of a drug defined the addiction it produced—not the kind of drug user. Thus, for example, the committee included minor tranquilizers in the section on "barbiturate-type dependence," writing that "by analogy, all agents which produce barbiturate-like sedation, because of the relief of anxiety, mental stress, etc., should produce some psychic dependence and . . . physical dependence when a sufficient concentration in the organism has been attained."[39]

The 1964 report from the Committee on Addiction-Producing Drugs dealt delicately with the policy implications of the new guidelines. "It must be emphasized," the authors explained, "that drug dependence is a general term that has been selected for its applicability to all types of drug abuse and thus carries no connotation of the degree of risk to public health or need for any or a particular type of drug control."[40] This was to some degree disingenuous, as researchers like Eddy had become increasingly convinced of the need for tighter federal oversight of prescription sedatives. The statement may have been more honest in its implication that, as lawmakers regulated new categories of drug dependence, they remained free to distinguish legally between acceptable and criminal addiction. Pharmacology might define drug *dependence*, the guidelines held, but did not necessarily define drug *addiction* or the policies designed to curb it. Pharmacological researchers, in other words, did not necessarily set out to overturn social (or legal) paradigms of addiction.

By 1965, ideas about addiction had changed to the extent that federal legislation finally became possible. The political rebirth of the disease model of addiction in the decade since the Boggs Act made it easier to imagine the possibility of addiction in respectable quarters. The nuanced pharmacological framework emerging from researchers like Isbell and Eddy made it easier to control prescription drugs without, as Anslinger had feared, "putting everything in the same compartment as morphine." The new focus on pharmacology also allowed a political compromise with professional and commercial groups. Calling for tighter controls on all sedative or stimulant drugs meant that all commercially competing drugs would be included, thus eliminating any potential commercial advantage. It also allowed Congress to delegate to the Food and Drug Administration the task of figuring out which particular drugs would fit the definition. This meant that physicians' groups and pharmaceutical industry lobbyists could spend their time battling FDA experts rather than lawmakers.

The legislative history of the 1965 law illustrates how the changing vocabulary of addiction opened up new possibilities for political action. After the 1955 defeat, the campaign to rein in prescription medicines had started up again in 1961, when Senator Thomas Dodd of Connecticut introduced a bill to regulate barbiturates and amphetamines. Each year, however, lobbyists from organized medicine and the pharmaceutical industry would successfully block the bill with the same tactics that had been successful in the 1955 Boggs Act hearings.[41]

In 1964 Dodd and his allies tried a new tactic: instead of trying to regulate individual drugs, they would try to regulate a class of drugs that shared a particular kind of pharmacological effect. Drawing from President Kennedy's Advisory Commission's 1963 *Report*, they defined these drugs as "psychotoxic": "any chemical substance capable of adducing mental effects which lead to abnormal behavior." In keeping with the *Report*'s categories, they included the minor tranquilizers as well as the barbiturates and amphetamines in this definition. Most such drugs, the *Report* acknowledged, had "legitimate medical use," but they could be abused when taken "for their psychotoxic effects alone and not as therapeutic media prescribed in the course of medical treatment." The commission therefore recommended that "all non-narcotic drugs capable of producing serious psychotoxic effects when abused be brought under strict control by federal statute."[42]

"Psychotoxic," of course, is nearly as vague a term as "dope," but the new language did nonetheless call for regrouping drugs on the basis of pharmacological action. The psychotoxic category would include all drugs producing certain identifiable effects regardless of their medical status. The pharmacology was still crude and made little sense—both the hallucinogen LSD and sedatives like the barbiturates belonged in the same group under its logic—but it was a step away from even more sweeping language arbitrarily denoting all illegal drugs addictive and all medicines safe or merely "habituating." Moreover, "psychotoxic" opened the way for political compromise by making it possible to talk about substances "capable of" rather than always and in every case producing toxic effects. This could support a system in which drugs—and medicines —were "psychotoxic" only if taken for the wrong reasons.

Again, the hearings were full of scare stories about truck accidents and "horrible crimes" increasing at "frightening rates," and everyone seemed to agree that something needed to be done. But once again the opposition prevailed. The American Medical Association called for solving the problem with "education and appropriate local and State laws." The Pharmaceutical Manufacturers Association agreed to tighter regulation but rejected the appellation "psychotoxic," saying that its broad meaning gave government a virtual blank check to regulate any drug, and that it unfairly stigmatized medically useful drugs. The association also opposed regulating drugs according to their "potential" for abuse rather than waiting for objectively witnessed and measured evidence of abuse.[43]

But the public pressure to do something, particularly about barbiturates, was increasing. Senator Dodd appeared on CBS news blaming "undercover opposition" for stalling his bill in the House, prompting a somber editorial by anchor Eric Sevareid. That same year famed reporter Walter Cronkite ran a special exposé on CBS news revealing just how easy it was to buy prescription sedatives without a prescription. His staff, pretending to be buyers for a fictitious export-import firm, had successfully bought enough barbiturates to make more than a million pills— worth up to $500,000 on the black market—for $600.[44]

In 1965 Dodd finally got his bill passed by revising key components that had drawn criticism. He abandoned "psychotoxic" in favor of language that more carefully followed recommendations Isbell had sent the previous year. As per Isbell's suggestions, the new bill would cover not just barbiturates and amphetamines but "any drug which contains any quantity of a substance which the Secretary [of Health, Education, and Welfare], after investigation, has found to have, and by regulation designates as having, a potential for abuse because of its depressant or stimulant effect on the central nervous system or its hallucinogenic effect." According to the final bill, the Food and Drug Administration (FDA) could categorize a drug as having a "potential for abuse" if it encountered evidence that "individuals are taking the drug . . . in amounts sufficient to create a hazard to their health or the safety of other individuals or of the community; or there is significant diversion of the drug or drugs containing such a substance from legitimate drug channels; or individuals are taking the drug . . . on their own initiative rather than on the basis of medical advice from a practitioner licensed by law to administer such drugs in the course of his professional practice."[45]

The threshold for proving such facts was apparently to be quite low, as the House of Representatives made clear in a report officially entered into the record alongside the new law:

> [A] drug's "potential for abuse" should be determined on the basis of its having been demonstrated to have such depressant or stimulant effect on the central nervous system as to make it reasonable to *assume* that there is a substantial potential for [abuse] . . . the Secretary of Health, Education, and Welfare should not be required to wait until a number of lives have been destroyed or substantial problems have already arisen before designating a drug as subject to controls of the bill.[46]

This message underscored the distance the new act had gone toward policing pharmacology rather than drug users. Substances were to be controlled based on their effect on the central nervous system, not on the crimes committed by abusers. If these crimes entered into consideration at all, it was only to prove that it was "reasonable to assume" that the "potential" for abuse was indeed present.

This new focus on a drug's potential for abuse provided a persuasive rationale for including Valium and other tranquilizers, which clearly produced "depressant effects." Indeed, the tranquilizers were a prime reason for that language in the first place. Nonetheless, the bill ultimately made no mention of Valium or any of its competitors. Despite the lobbying of barbiturate manufacturers, who had pushed for explicit inclusion in the bill of competitive drugs like the tranquilizers, Senator Dodd, following Isbell's advice, chose not to list other regulated drugs separately, hoping thereby to avoid having to battle lobbyists from each drug company. For too many years similar bills had died in the face of just such opposition. (The executive vice president of Roche, for example, was perfectly willing to support the law so long as Roche's own flagship drugs Librium and Valium were not named in it.)[47] Instead, the law called on the FDA to determine if drugs other than barbiturates and amphetamines met the criteria for "having a potential for abuse." To placate barbiturate manufacturers, the House report specifically named minor tranquilizers as among the drugs "already causing serious problems" requiring study and classification. The Senate report included assurances from the Department of Health, Education, and Welfare that the agency would act quickly to prevent "unwarranted competitive disadvantages" to makers of barbiturates.[48]

The Drug Abuse Control Amendments of 1965 required strict record-keeping by manufacturers, distributors, pharmacies, and dispensing physicians, and limited the number of refills on a single prescription to a maximum of five within a limit of six months. (Under previous federal law, prescriptions could be refilled indefinitely with a physician's permission.) This legislation was able to become law, however, in part because of what it did not do. It did not label the medicines "addictive," but simply as having the "potential for abuse." Its primary focus was to rein in nonmedical use, either through illicit trafficking or individual use beyond a physician's recommendation. Physicians could no longer delegate the authority to refill to patients but could otherwise prescribe as they saw fit. Moreover, possession for "personal" or even "family" use continued to be perfectly legal, even if the drugs had been obtained illegally. Perhaps most important, the various provisions would be enforced by a new Bureau of Drug Abuse Control in the FDA, not by the Narcotics Bureau.[49]

Despite these limits, the strict record-keeping and official acknowledgment of risk did establish an important basis for recognizing addiction in unlikely places. Rising sympathy for addicts and increasing awareness of drug actions had helped produce a shift toward emphasizing the role of drugs rather than character flaws in addiction, but the shift was a partial and highly contested one.

Addictive Medicines or Susceptible Persons?

The Drug Abuse Control Amendments had thrown the legal fate of the minor tranquilizers into the hands of the Food and Drug Administration, a compromise the bill's sponsors (and the barbiturate manufacturers) accepted only because of an explicit promise from FDA officials that they would immediately add the minor tranquilizers and other drugs to the "potential for abuse" category. True to its word, the FDA proposed regulations in early 1966 that incorporated meprobamate (Miltown) and the benzodiazepines (Librium and Valium), along with a short list of other drugs.[50]

The inclusion of the tranquilizers was hardly the *fait accompli* legislators thought it was, however. Carter Products, maker of the Miltown brand of meprobamate, and Roche Laboratories, maker of Librium and Valium, immediately challenged the new regulations, initiating years of legal wrangling in the FDA and the court system to determine whether they met the criteria for "abuse potential" as defined in the new law.[51] Were tranquilizers potentially addictive? Had this potential actually translated into real abuse (nonmedical use) and actual addicts, and did or should that matter before the law? And if one kind of tranquilizer were scheduled, should newer tranquilizers of the same sort be scheduled as well based on their pharmacology, or should the government wait for evidence of potential for (or even reality of) abuse? The tranquilizers became the test case of pharmacological reasoning for two reasons: first, barbiturates had been specifically named in the original bill, and second, unlike the barbiturates, the tranquilizers had little or no reputation for illegal use or abuse. In the 1960s, at least, their risk seemed almost entirely potential.

One reason Carter and Roche were so successful in slowing the political process was the ambiguous and contested state of medical knowledge about Miltown and Valium. Antidrug reformers like Nathan Eddy and Harris Isbell may have been influential, but they were hardly the only people to claim "expertise" over the minor tranquilizers. Pharmaceutical companies and many practicing physicians also circulated ideas and shaped the medical landscape. To understand the post-1965 legal and political struggles, we need first to examine these prevailing medical ideas about addiction and the minor tranquilizers in the 1950s and 1960s.

Not surprisingly, drug companies and many physicians were not convinced that these widely used medicines could actually be addictive. At most, they allowed, addiction might result when "dependence-prone" people refused to follow physicians' instructions. But evidence of physical dependence on high doses did not, to their minds, establish a parallel with illegal drugs. Unlike illegal drugs, tranquilizers could be used safely by ordinary people; the medicines did not turn people into junkies.

Rather than outlaw the medicines, the trick was to strengthen physicians' authority so as to keep the drugs out of the hands of junkies, or people predisposed to become junkies.

What was known or suspected about dependence on tranquilizers provided fertile territory for such debates. Unlike opiates, which had provided early addiction researchers with a clear and relatively unvarying picture of dependence, tranquilizers resisted easy characterization. Opiates and other clearly addictive drugs offered immediate pleasurable reinforcement; Valium's effects, on the other hand, were slow to take hold and could be quite subtle at low doses. Moreover, dependence could take a long time to develop—by some accounts months or even years at low doses. And while withdrawal from high doses may have been obvious (seizures, vomiting, temporary psychosis, and other symptoms were difficult to miss), withdrawal from lower therapeutic doses resembled nothing so much as the symptoms of anxiety for which the drug had probably been prescribed in the first place: irritability, restlessness, inability to concentrate, panic attacks, tremors, sweating, insomnia, and so forth. Intermediate withdrawal effects of muscle twitching, nausea, aches and pains, and ultra-sensitivity did not fully clarify the picture.[52]

When Carter Products first released Miltown, even these ambiguous possibilities were unambiguously absent. Early ads baldly promised that the drug was "not habit forming"—an important claim in 1955, when barbiturates (potential competitors) were under heavy fire in Congress. A 1956 ad, published after early reports of addiction by Lemere and others, still claimed Miltown was "well suited for prolonged therapy" because it was "nonaddictive."[53] A "clinical bulletin" mailed to physicians in early 1957 followed similar lines, stating that "habituation does not follow the use of Miltown" and that "withdrawal symptoms have been completely absent"; the drug was "ideal . . . for repeated use, as in premenstrual tension."[54] Even after the FDA insisted that Carter take reports of dependence into account, the new Miltown brochure simply advised that "possible habit formation" might take place "in those patients prone to excessive self-medication, such as alcoholics and severe neurotic personalities." Acrobatically avoiding the language of addiction and withdrawal, Carter recommended gradually reducing doses rather than stopping drug treatment abruptly because "the withdrawal of a 'crutch' may precipitate an anxiety reaction of greater proportions than that for which the drug was prescribed." Only upon FDA insistence did Carter revise the language to warn that "sudden withdrawal may precipitate withdrawal reactions," even as the general warning—to be careful of "excessive and prolonged use in susceptible persons"—remained the same.[55]

Unlike Carter Products, most physicians acknowledged the reality of withdrawal seizures and other symptoms resulting from long-term massive doses of Miltown. As we saw, both Essig and Eddy confirmed Lemere's early suspicions in separate articles for *JAMA* in 1957. Other researchers continued to add new evidence. Typical was a 1958 study published in the *New England Journal of Medicine*, which detailed the "insomnia, vomiting, tremors, muscle twitching, overt anxiety, anorexia and ataxia" of nearly fifty patients switched from Miltown to placebo without their knowledge. Eight patients "showed a picture of hallucinosis with marked anxiety and tremors," while three had "grand-mal seizures."[56] By 1960 the *Medical Letter on Drugs and Therapeutics* reported "increasing evidence" that "addiction and withdrawal symptoms, convulsions, coma, psychotic behavior, and even death" could follow prolonged use of high doses of Miltown.[57]

But as Carter Products was quick to point out, no one was supposed to be taking such high doses. And addiction to lower therapeutic doses— the amounts prescribed by physicians—remained a far more ambiguous proposition. Many physicians reasoned that milder "jitteriness" upon withdrawal from such doses simply indicated a return of the patient's initial nervous condition. Indeed, Frederick Lemere, the psychiatrist whose *JAMA* article first sounded the alarm about Miltown addiction, had noticed a problem only when patients taken off the drug kept making remarks like, "I guess the medicine was doing some good after all because when I stopped it, I certainly felt nervous."[58] An Albany psychiatrist, critiquing Lemere's report, announced that he himself had never observed withdrawal symptoms, although "sometimes there was a return of the symptoms existing prior to therapy." "It would be a pity," he concluded, "if a drug as useful as [Miltown] . . . should acquire an undeserved reputation for habituation."[59] The early reference guide *Drugs of Choice, 1958–1959* found "little evidence of habituation to [Miltown] and no real tolerance . . . even though many individuals are reluctant to discontinue its use."[60] Another study published in *JAMA* in 1958 reported "no true habituation" in Oklahoma prison inmates tested with

Miltown, although when the subjects were withdrawn from the drug, the "original symptoms of which [they] had complained returned with shocking suddenness." (The study also noted—but did not dwell upon— two "typical *grand mal* seizures.")[61]

If it was hard to believe that therapeutic doses of Miltown could be addictive, it was even harder to accept such damaging news about the newer and even more popular Librium and Valium, about which much less evidence was initially available. But as early as 1961, the year Librium first became widely available, researchers at Stanford University reported symptoms "definite and consistent with a withdrawal reaction" among patients who suddenly stopped taking high doses of the drug. Two years later, the group published a similar report about Valium. These studies had been designed specifically to counter skeptics, focusing on the appearance of new symptoms during withdrawal whose intensity had fluctuated precisely in accordance with levels of the drug measured in the bloodstream.[62]

Careful studies like these from Stanford were rare, however, especially given the important role of pharmaceutical companies in funding drug studies. Tranquilizer manufacturers, not surprisingly, were not inclined to spend money in an effort to prove their own best-selling products unsafe. A representative of Roche Pharmaceuticals later admitted to a Senate subcommittee, for example, that Roche had not funded a single study of Valium's potential for addiction. "You asked for a list of studies which 'we did or may have supported' on dependence and long-term effects of Valium," wrote Roche's assistant vice president for public affairs. "As far as I have been able to find out, Roche did not actively conduct or support such studies." This was a striking admission when one considers how actively Roche worked to persuade regulators that the drug was nonaddictive.[63]

The result of such funding scarcity was predictable: poor-quality, uncontrolled studies that left the question of addiction unresolved. A 1976 survey of existing literature on Valium, for example, found that literally dozens of clinical trials throughout the years had reported no habituation to the drug, but the vast majority of them had failed to establish protocols for measuring withdrawal, or even to identify any criteria by which to determine its presence or absence.[64] Typical of this methodological looseness were the almost offhand findings of a Texas gynecologist, who reported in 1962 that of one hundred women tested with Librium, "A number . . . asked that they be put back on the medication after it had been discontinued for a time, but this was due to continued or renewed stresses rather than to any physiologic demand for the drug. Certainly, its addictive qualities must be very low or nonexistent at usual therapeutic dosage levels."[65]

The difficulty of proving therapeutic-level addiction strengthened a second argument used to defend the tranquilizers: that they were safe unless used by "dependence prone" people—the kind who might, for example, decide to use more than the physician had advised. Such reasoning seemed a perfect way to explain the apparent rarity of minor tranquilizer addiction. In a second critique of Lemere's initial *JAMA* report, a Boston surgeon noted that ten of Lemere's thirteen patients had been, or still were, alcoholics. Such individuals "present a special case as far as addiction and excessive self-medication are concerned," the surgeon argued, and a report of "possible habit-forming properties of [Miltown] in a few alcoholics" hardly represented a gold standard for scientific evidence. A "fairer" test would be to give the drug to "normal persons . . . in a time of stress." He himself had undertaken such a test and had never seen a case of what he called "genuine" withdrawal in over 250 "normal" patients treated with the drug for postoperative stress.[66]

The surgeon's criticism was difficult to refute. Most documented instances of minor tranquilizer addiction did indeed involve people with a history of alcohol or drug abuse.[67] Even when this was not the case, most minor tranquilizer studies were (understandably) not performed on "normal" or fully healthy people, but rather on people suffering from nervous conditions of one sort or another—people, in other words, who could easily be seen as predisposed to drug dependency. In the 1958 Oklahoma prison study cited earlier, for example, the author described the subject population as one "in which a high incidence of susceptibility might normally be expected" since "most of the subjects were basically unstable, unhappy, frustrated individuals." Although the study found no "true" habituation, it warned that when treatment was suddenly stopped, "a small percentage of patients, especially those with central nervous system damage or serious mental, emotional, or characterological abnormalities, may react in an undesirable manner."[68]

Since almost all studies of tranquilizer addicts in these early years involved men, not women, the "dependence-prone" argument was also bolstered by popular representations of male tranquilizer takers. For a man to take a tranquilizer was already to cast suspicion on his strength, virility, and overall masculine identity. This sort of inadequate masculinity was at the heart of the increasingly psychological explanations for addiction emanating both from groups like Alcoholics Anonymous and addiction treatment specialists in the 1950s and early 1960s.[69]

Reports of addiction to tranquilizers often turned to such explanations. One 1957 case history in the *New England Journal of Medicine*, for example, described a Miltown addict as "a very inadequate and dependent personality with the ability to become easily dependent on anything he can swallow."[70] Another physician, writing in *Medical Clinics of North America*, explained that addiction to the drug "usually occurs in extremely dependent, emotionally immature persons who will grasp upon any means to maintain themselves free of their inner tension."[71] A group of Kentucky researchers offered a similar description of Valium addicts in *JAMA* in 1966. "As a group," they noted, addicted patients "were decidedly 'dependence-prone' in the sense that they were chronically anxious and dependent, and were bound to seek relief from these symptoms by excessive self-medication with the drugs in question once they had discovered their sedative properties as a result of prescriptions by their physicians."[72] Even apparently normal individuals who became addicted to minor tranquilizers, a Mayo Clinic researcher discovered, were in reality dependence-prone, their "appearance of strength and good adjustment . . . only a veneer."[73]

Ironically, the same litany of personal and emotional problems that supposedly made someone predisposed to addiction was also perfectly constituted to bring someone into therapy with a tranquilizer. That this, in theory, would make them "dependence-prone," and thus at risk of becoming dependent on tranquilizers, seems to have remained unexplored.

The argument that addicts' own personal flaws were to blame for their addiction was not restricted to tranquilizers, of course. As the AMA's Committee on Alcoholism and Addiction and the Council on Mental Health declared jointly in 1965, "Drug dependence is a medical syndrome, a symptom complex, and almost always reflects some form of underlying mental disorder which has preceded and predisposed the patient to the development of drug abuse."[74] Indeed, the importance of psychiatrists in medicine's push to reclaim authority over addiction virtually ensured that drug users' mental health would continue to play a key role in defining the problem. Leading addiction experts like Lawrence Kolb in fact had long championed the notion that psychological characteristics were the key element of addiction. In the antinarcotics campaign, these psychological problems had melded seamlessly with moral condemnations of innate immorality and criminality. Defenders of prescription medicines in the 1960s, however, enjoyed a more sympathetic context in which drug problems conjured visions of weakness and illness rather than viciousness and depravity. Tranquilizer addicts were to blame for their dependence but were portrayed as more pathetic than dangerous. Virtually no studies suggested the dangerous and immoral junkie profile that had emerged from addiction research on heroin in the 1920s.

Among the most enthusiastic adherents of the "dependence-prone" argument were drug manufacturers. The *Physician's Desk Reference* (*PDR*), for example, a standard drug reference tome whose entries were written by pharmaceutical companies, strongly supported this argument. The 1959 *PDR* entry for Equanil stated that "excessive and prolonged use in susceptible persons, such as alcoholics, former addicts and severe psychoneurotics, can result in dependence on the drugs." Miltown's entry for that year advised caution in prescribing for "patients with a known propensity for taking excessive quantities of drugs," because "excessive and prolonged use in susceptible persons has been reported to result in dependence." Early *PDR* entries for Librium and Valium also warned about prescribing to "individuals known to be addiction-prone, or whose history suggests they may increase the dosage on their own initiative."[75]

Not everyone believed in the utility of the "addiction-prone personality" concept. Carl Essig, for example, argued that, in practice, it was difficult to identify addiction-prone individuals or to predict which patients were most likely to abuse sedatives and minor tranquilizers. "The usual safeguards," he advised in 1964, "are to limit the amount of drug prescribed and prevent refilling of the prescription."[76] "Addiction prone," in short, was often a retroactive diagnosis, and thus an unreliable guide for physicians attempting to decide how to prescribe potentially abusable drugs. For practical purposes, Essig suggested, it made more sense to handle drugs according to their potential to be abused, rather than trying to differentiate between different kinds of drug users—exactly the strategy Congress had finally adopted in the 1965 Drug Abuse Control Amendments.

"Not the Cases We Intend to Prosecute": Valium on Trial

Experts like Essig may have helped bring barbiturates and amphetamines under new federal regulations, but as we have already noted, the newer drugs Miltown and Valium would be included under the 1965 law only if the Food and Drug Administration determined that they had a "potential for abuse." This was no sure thing. As we shall see, the partial and contested shifts in thinking about addiction and medicines led to a partial and contested acceptance of tranquilizers as drugs of potential abuse. They were included in the law, marking a victory for the pharmacologically based "disease model" of addiction, but they were placed in a special category that attempted to maintain lines separating them from dope.

During the FDA hearings, parades of medical and pharmaceutical industry experts battled over the meanings of addiction and the qualities of Valium and the tranquilizers. Indignant psychopharmacologists cited clinical trials and statistics to defend their discoveries as safe and non-addictive. Ordinary physicians insisted that they had never seen an instance of tranquilizer addiction. Drug company representatives continually questioned the language and implications of proposed decisions and guidelines. Meanwhile, on the other side, addiction specialists (Nathan Eddy was ubiquitous at the hearings) presented their own clinical trials, statistics, and patients' testimony in a determined effort to prove the urgency and seriousness of the risk. The FDA's prescription law enforcement unit told of raids on pharmacies and arrests of pharmacists and patients for buying and selling the drugs without proper prescriptions. Even Essig's poor Miltown-addicted dogs made an appearance: Eddy showed a film of their untimely demise to the hearing examiner and, much to Carter Products' dismay, leaked it to the press.[77]

Ultimately, the FDA examiners in both cases sided with the government, finding a potential for abuse in Miltown and Equanil in 1967 and in Librium and Valium in 1969.[78] In making these decisions, both hearing examiners relied explicitly on the new logic of pharmacological potential. Carter Products, for example, initially protested that "isolated or occasional" instances of Miltown abuse resulted not from the drug's depressant effect on the central nervous system but rather from a predisposition in some users of these drugs. Frank Berger, Miltown's discoverer and Carter's research director, called addicts "abnormal" and explained that their addiction "is not due to the drug, but is caused by their abnormal personality." The fact that their "compulsion" was for Miltown, he insisted, was merely "coincidental." Carl Essig also admitted that narcotics addicts and alcoholics were the most liable to abuse Miltown.[79] The hearing examiner, however, held to the pharmacological standard allowed by the law: "What effects this drug does produce," he concluded, "are due to its depressant effect on the central nervous system." Even if other factors came into play, this fact combined with documented instances of actual addiction were enough to legally establish the potential for abuse.[80]

That pharmacological characteristics alone were nearly sufficient to indict a drug became even clearer in the Librium and Valium hearings. Responding to physicians who claimed they had seen no dependence on Librium or Valium in "extensive clinical use over many years," the hearing examiner bluntly declared that such testimony "does not rebut the evidence introduced by the Government." (This evidence had included medical accounts of addiction and, even more damning, reports from local police and public health agencies that indicated abuse and widespread criminal diversion of the drugs from pharmacies: over two hundred thousand Librium and Valium pills went missing from just three pharmacies tracked in 1968. Such information, the researchers noted, had not previously been available because studies tended to focus on controlled substances—which Librium and Valium, of course, were not.) The government had shown that the tranquilizers could produce physical dependence, even if they did so very rarely, and thus had met the requirements of the law. As the hearing examiner's summary statement declared, "The evidence establishes that it is possible to develop tolerance to Librium and Valium, but that it has not been frequently observed or reported."[81]

Interestingly, in neither case did the examiners completely reject the idea that some people had a predisposition to addiction. After citing Hollister's 1961 study as evidence that Librium could produce physical dependence, for example, the hearing examiner proceeded to explain that Librium addicts "usually had experienced difficulty with similar drugs before . . . [These] 'dependent personalities,' as they have been characterized, look for chemical solutions to a variety of problems."[82] Miltown's hearing examiner similarly acknowledged that the minor tranquilizer was most attractive to "individuals who have a propensity for becoming dependent on certain types of drugs."[83] The examiners were also respectful of other common objections to regulating the tranquilizers. If pharmacology was such an important basis of addiction, one asked, why did so very few people exposed to depressant or stimulating drugs become addicts?[84] Moreover, how could or should a law based on pharmacological potential account for addicts who did not act like drug abusers, such as postsurgical morphine addicts who willingly and successfully went off the drug?[85]

While they rehearsed the medical profession's lack of consensus on these issues, however, the FDA ruling ultimately finessed them. Following the pragmatic reasoning articulated by Carl Essig, the hearing examiners based their decisions on a narrow reading of the law and left speculation about such broader questions to other venues.

Carter Products and Roche Pharmaceuticals both independently appealed the FDA's decisions, catapulting the process into the legal system. Carter lost its appeals in 1969, and in that year Miltown finally came under the 1965 law. Roche, on the other hand, actually won its case on a technicality in 1973. Librium and Valium only came under government regulation in 1975, after a surprise out-of-court settlement between Roche and the newly-formed Drug Enforcement Administration.[86]

This meant that the drugs' legal fate was still in the air in 1970, when the stakes grew considerably higher. In 1968 the FDA's enforcement arm, the Bureau of Drug Abuse Control, had been merged with the Federal Bureau of Narcotics. Both had joined the Justice Department in a new Bureau of Narcotics and Dangerous Drugs—a single enforcement entity that oversaw all kinds of regulated drugs, regardless of the nuances of how they had come to be regulated. In 1970 Congress began to debate a new proposal from President Richard M. Nixon that would fold the bureau into a new entity, the Drug Enforcement Administration, with even stronger police powers.[87]

While the status of Librium and Valium was a minor issue in these debates, the subject did come up with some regularity, and Roche in particular lobbied hard to ensure that the new legislation did not preemptively regulate its flagship drugs before the legal appeals had been exhausted.

Roche's efforts focused on one key question left unanswered by the Drug Abuse Control Amendments of 1965 and the FDA: If tranquilizers had the pharmacological *potential* to produce addiction, should they therefore be controlled like any other addictive substance? Or was the occasional tranquilizer addict so harmless as not to be a public menace? These two positions distilled the debate over addiction to its most fundamental terms: Was the problem drugs, or the kinds of people who used them? Were there different kinds of addiction, and if so, how should such distinctions be incorporated into a consistent set of regulations?

The initial draft of the 1970 bill covered all drugs of abuse with a single statute by creating a "schedule" of four classes of "controlled substances." Ranked by medical usefulness and potential for abuse, the classes were modeled after the latest pronouncements of the World Health Organization and seemed a victory for pharmacological specificity: based on their potential to foster abuse, the different schedules would provide for different levels of control.

Schedule I included heroin, marijuana, LSD, and other drugs considered to have a "high potential for abuse," "no currently accepted medical use," and "a lack of accepted safety for use" even under medical supervision. Schedule II included cocaine, morphine and some other opiates, and other drugs determined to have a "high potential for abuse," accepted medical use, and a risk of "severe psychological or physical dependence." Schedule III drugs, including the amphetamines, the barbiturates, and Miltown, had a lower potential for abuse than Schedules I and II, "accepted medical use," and a potential for "moderate or low physical dependence or high psychological dependence." Schedule IV featured combination drugs that included a narcotic and had a "low potential for abuse" and were limited in their potential for dependence because of the interference of the other drug in the combination.[88]

Although Librium and Valium were not mentioned by name in the bill, Miltown had already lost its case and was thus listed in the proposal in Schedule III. Librium and Valium would presumably join it there after Roche's legal appeals were over. The drugs' many defenders, however, were quick to point out that they considered this to be a mistake in the making. "It is difficult to fathom," testified Neil Chayet, an expert in legal medicine and member of NIMH's Narcotic Addiction and Drug Abuse Review Committee who testified for the pharmaceutical industry, "how drugs such as . . . [Librium or Valium] can be classed in the same schedule with methamphetamine, one of the most abused and deadly of substances." Chayet argued that a drug's pharmacological potential to produce dependence had to be placed within the context of its potential to produce real social harm. "A far more rational criteria for control," he recommended, "would be one which stressed the element of danger—that is, the danger which a substance poses for the user and the impact of the substance on the individual and on the public health of this nation."[89]

Dr. Jonathan Cole, the first head of the National Institute of Health's Psychopharmacology Service Center, also emphasized the difference between physical dependence and the social problem of drug addiction. Miltown, Valium, and Librium "are not drugs of abuse in the usual sense," he explained. "People become addicted to (dependent on) them by taking too much of what the doctor gave them, rather than by going out and buying them on the street from pushers." Declaring the bill's terminology "a morass," Cole again stressed that "medical addicts" were "a different . . . class from those who buy heroin on the street."[90] Such rhetoric echoed testimony the previous year from Dr. Henry Brill, director of New York's largest psychiatric hospital. Medical drug abuse, he explained, "is mature men and women, solitary, diffused through the social structure, and has no connection to antisocial behavior," while nonmedical abuse involved "young male pleasure-seekers, a group of people who are socially contagious and are a threat to others."[91]

In a strictly literal sense, of course, Cole was quite right: minor tranquilizer users enjoyed a more affluent social background than the average heroin addict. But what did this say about the drug itself? Where did such social realities belong in drug policy? Chayet, Cole, and other critics rejected the reasoning that equated physical dependence with the social problem of drug addiction. Dependence, they maintained, was a medical condition and became true addiction only when it began to cause social harm. As psychiatrist and prominent psychopharmacologist Frank Ayd pointed out, "There is no evidence that a serious problem has been created for society by the infinitesimally few abusers of these compounds."

Daringly, Ayd even speculated that some tranquilizer dependence might be perfectly understandable "self-medication" by patients whose physicians were reluctant to prescribe high enough doses. This was an interesting analysis, but one that Ayd tellingly did not go out of his way to extend to people dependent on other, "harder" drugs like narcotics.[92] Such reasoning hearkened back to earlier efforts to distinguish between acceptable and dangerous forms of drug dependency—the "habituation" versus "addiction" split. It kept a space open for the traditional predisposition argument by continuing to locate the true dangers of addiction in the behavior of users rather than drug pharmacology.

While such reasoning certainly had some merit, it was also highly susceptible to the kinds of class and ethnic stereotyping that had long shaped perceptions of drug use in America. Was tranquilizer addiction a relatively mild and nonthreatening matter? Or were tranquilizers primarily used (and thus primarily abused) by nonthreatening populations —white, respectable, largely female—who were able to maintain their habits without, for the most part, breaking the law or actively participating in criminal urban subcultures? For some, such distinctions simply did not matter. Tranquilizers were not a social problem; narcotics were. This apparently pragmatic distinction pleased those already inclined to support Valium and its cousins. It also, however, continued to blur the difference between social and pharmacological reasoning.

Adding another wrinkle to an already complex situation was the fact that even "respectable" prescription drug addicts committed their fair share of crimes. Forged prescriptions, borrowed or stolen pills, and multiple physician visits were everyday life activities for many medical addicts.[93] Those crimes, however, were hardly the stuff of the "dope fiend menace" and appear to have been taken rather lightly by lawmakers in 1970. One senator, for example, asked how the new law would handle what he considered a typical instance of tranquilizer abuse: "A doctor prescribes Librium for some nervous lady. She is making her husband nervous, so he takes a couple of pills to work with him and has them in his possession at lunchtime. You catch him with possession. Does he come under this act?" The DEA's chief counsel reassured the worried senator: "Technically, yes, but . . . these are not the cases we intend to prosecute."[94]

This atmosphere of uncertainty about the risks of Valium presented an opening for Roche, whose lobbyists proposed during the 1970 hearings that Congress add a new schedule for the benzodiazepines. This schedule would impose the same regulatory controls as Schedule III, but it would "be distinguished by a different label symbol for the drugs it covered" and would make first-offense violations a misdemeanor rather than a felony.[95] Roche buttressed its claim by referring to the 1970 guidelines of the WHO Committee on Addiction-Producing Drugs, whose four schedules had also put tranquilizers in a less severe category.[96] Significantly, the company also appended to its official statement eleven pages of letters solicited from police departments across the nation attesting that abuse of Librium and Valium was "definitely not a problem" in their city.[97] The strategy worked, apparently. When the 1970 Drug Abuse Control bill became law, it had not four but five schedules, including what critics called the "Roche schedule" (IV) for minor tranquilizers —even though Roche had successfully kept its own minor tranquilizers, Librium and Valium, out of the bill altogether.[98]

Conclusion

The medical and political debates over the addictiveness of the minor tranquilizers were an important chapter in the history of America's evolving understanding of addiction. The mere fact that these prescription medicines had come under such close scrutiny revealed a significant breakdown in the Anslinger-era dichotomy between medicine and dope, a central pillar of the reasoning that held addiction to be a criminal act rather than a disease of exposure. New divisions had arisen within the once solid antidrug coalition, opening unsettling questions about the relationship between pharmacology and the social problem of drug abuse. The resulting debates produced significant accomplishments in the black-and-white world of American drug politics: they acknowledged the risks of prescription medicines and at least potentially resurrected the long-forgotten figure of the medical addict to stand alongside junkies in the cultural iconography of addiction.

As the Valium saga shows, however, renegotiating addiction did not take place in a social vacuum. Pharmaceutical companies and many physicians exploited addiction's inner ambiguities and contradictions to absolve their favored drugs—a powerful advocacy from which other, already illegal drugs did not benefit. In many cases this advocacy drew directly on older logic that equated true addiction with deviance, criminality, and other elements of the junkie paradigm.

Take, for example, the claim that Valium itself was essentially innocent even if people became dependent on it, because responsibility lay with the drug user. Similar arguments had not protected opiates in the 1910s and 1920s, because the behaviors central to maintaining heroin addiction had been demonized during the era's antivice crusades. Junkies thus appeared to be not just flawed individuals but depraved and vicious criminals. Valium addicts, on the other hand, could be persuasively described as weak rather than demonic, needing laws to protect rather than punish them. Why? In part the answer lies in demographics: Valium users were largely white, middle class, and female. But even this fact emerged from the broader dynamics of postwar commercial medicine, within which respectable Americans were legally and affordably able to maintain a supply of their drug in impeccably legitimate locales like pharmacies—and were openly advised by medical authorities and advertisers to do so. This system made it dramatically less likely for Valium users to even consider the possibility of committing a crime to get their drug (and, for that matter, much easier to commit a crime if they did). In short, the question of whether tranquilizers could produce junkies was not really tested—a condition made possible by commercial medicine, and in turn usefully exploited to defend commercial medicines.

Valium's defenders looked at this difference in patterns of drug use and attributed it to Valium itself. By its nature the drug simply did not produce junkies, they argued, although junkies might unwillingly turn to it in the absence of their favored drugs. This elision of social and pharmacological categories helped sustain distinctions between the street and the medicine cabinet even as new regulations settled around tranquilizers. And by certain lights, it made pragmatic (and strategic) sense. After all, no one was claiming that an army of Valium junkies lurked around the corner. But it also meant accepting the vast and complicated social substructure that supported such divisions, cloaking them in the authoritative vocabulary of medical—and, of course, legal—difference.[99]

In the short term, such strategies appeared to pay off for Roche and others opposed to regulating Valium. The nation's most popular drug had been kept off the schedule of controlled substances on the manufacturer's terms, and a favored spot had been readied for it in case that changed in the future. But this achievement ultimately proved the downfall of Valium and the end of the minor tranquilizer era. Ironically, this unanticipated decade's-end decline resulted largely from the very social myths about addiction that Roche and others had so successfully used to defend their medicines. Valium's victory was a nuanced one, depending upon careful distinctions between types of drug dependence and types of addicts. In the popular arena, however, addiction was not a nuanced affair. After many decades of drug demonology, it had become a simple morality play with fiends, villains, and victims—something that Valium's manufacturers and prescribers learned to their dismay as they were unwillingly cast as the new pushers in a widespread panic over Valium addiction in the late 1970s.

CHAPTER FOUR

The Valium Panic

IN 1978, THE *FDA CONSUMER* described a new kind of junkie:

> The smartly dressed woman driving a sleek, late model car could be
> the envy of her neighbors. She has a loving husband, bright children, a
> beautiful home in the suburbs, and apparently no cares in the world.
> Except one. This woman is a junkie. She is not the kind of junkie one
> thinks of in terms of long-haired "hippies," counter-culture street peo-
> ple, pushers, and illicit drugs. She is dependent on legal drugs, the
> kind prescribed for her by a physician.[1]

As the article soon made clear, the "smartly dressed junkie" was not alone
in her addiction to Valium, the "new white-collar aspirin." Former first
lady Betty Ford became the most famous victim of "prescribed addiction"
when she admitted her dependence on Valium and alcohol in 1978, but
even before this, hair-raising stories like "Valium—The Pill You Love
Can Turn on You" abounded in popular magazines, in local newspapers,
and on television shows like *60 Minutes*.[2] Reclusive billionaire Howard
Hughes, famed coma victim Karen Ann Quinlan, and country music
singer Loretta Lynn had all reputedly succumbed. Barbara Gordon, television producer and author of a best-selling memoir about her addiction
to Valium (*I'm Dancing As Fast As I Can*), told *People* magazine of a
spreading fear of the drug: "I've been getting calls from women who are
panicked. 'I take two, I take four, I take six . . . Do you think I'm addicted?' "[3] No fewer than three congressional investigations of Valium
and similar drugs grabbed headlines. At one of these, a physician warned
that "if we were to stop abruptly the availability of this drug, our country
would be in an epidemic of anxious, neurotic, psychotic, trembling citizens in acute withdrawal."[4]

Most drug experts were dubious. As we saw in chapter 3, even during
the decade-long debate over regulating Valium, not even antidrug crusaders had claimed that addiction to the tranquilizer was an actual,

existing problem. Valium had been "scheduled" by the Drug Enforcement Administration because of its pharmacological potential, with the explicit caveat that actual abuse had rarely been observed and was not anticipated. Most of the eminent figures in psychopharmacology continued to defend their creations, dismissing fears of widespread addiction as sensational nonsense.[5] Even the epidemiologists whose surveys of national drug use fueled the panic argued that Valium and other mood medicines were actually *under*prescribed—in part owing to the irresponsible media.[6]

By the 1980s, when the intensity had waned, scholars questioned why popular hysteria had unfairly maligned a safe and valuable medicine ("Where are all the tranquilizer junkies?" asked the *Journal of the American Medical Association*).[7] The consequences, they pointed out, were quite real. National surveys showed that Americans' perceptions of the onetime "wonder drug" had darkened, and use of Valium had declined correspondingly.[8] By the 1980s new medicines like the antidepressant Prozac and the antianxiety drug Xanax (a pharmacological cousin of Valium) took Valium's place as America's leading psychiatric medicines. Japan saw no Valium panic and experienced no such shift to antidepressants.[9]

This was hardly America's first drug panic, or the first to be decried as exaggerated and unnecessary. Sensationalist panics were a longstanding American tradition, stretching from the temperance movement to the twentieth century's drug wars. And yet Valium's defenders were right to see something new in it. Past scares had targeted drugs associated with marginal populations like immigrants, nonwhites, or the urban poor. Indeed, antidrug rhetoric had long been an important tool for white middle-class cultural warriors looking to dramatize the threat posed by such "dangerous" classes to "our" society. The Valium panic, on the other hand, involved a quintessentially middle-class medicine prescribed legally by reputable physicians for their respectable patients and popularly recognized (with the help of ubiquitous marketing) as an entrenched part of life in the comfortable classes, especially for women. Bogeymen from the dangerous classes had no role in this drama. And the usual antidrug coalition of government, middle-class cultural crusaders, and medical authorities were all openly involved in the Valium trade.

The possibility of Valium addiction attracted a new coalition of activists, especially certain segments of the diverse second-wave feminist movement. They redeployed the powerful cultural tools of the antidrug genre for their own agendas, revising classic drug-scare narratives to sensationalize addiction among affluent white women as a central symbol of sexism and its consequences. Echoing Betty Friedan's early tranquilizer critiques, they held up liberation from "mother's little helper" as an archetypal story of emancipation through feminism. These stories became attached to Valium and remain fixed in popular memory about the drug even now, decades later.

This was, in many respects, a remarkable campaign. It earned new audiences for feminist political messages in mass-market media such as *Good Housekeeping*, *Redbook*, *Vogue*, and other places not usually considered feminist bulwarks. Its politically charged sympathy toward addicts challenged the punitive logic of the twentieth century's war against drugs. And it changed popular attitudes about Valium, helping reduce use of the drug to such an extent that we may never know whether the "epidemic" truly existed. The panic itself, in other words, may help explain the mysterious absence of Valium junkies in the 1980s.

However emancipatory and effective feminists' "use" of Valium was, however, their efforts did not ultimately break free of the race and class politics of the antidrug tradition. Valium crusaders came from the ranks of feminists most interested in galvanizing white middle-class women—the cultural quarters where the tranquilizer was most familiar. Part of their success came from their willingness to trade on assumptions about these women's essential innocence in a way that excluded—and even reified—the dangerous classes as a different sort of drug user. This was hardly a surprising approach given the campaign's goals. Anti-Valium crusaders did not set out to engender a broad-based reorientation of the war against drugs. And yet their success in reworking the antidrug tradition for their own ends challenges us to imagine exactly that: a war against drugs launched as a civil rights campaign, challenging rather than reinforcing cultural stereotypes.

This chapter examines the Valium panic as a window into three related dynamics. First, it investigates how activists could utilize commercial medicines to strengthen a cultural and political, not just psychological, identity. In an age of social movements organized in part around new political conceptions of identity, Valium became an arena for significant contestation. Second, it explores the historical and cultural connections between medicine-cabinet and street drug cultures. The Valium saga sheds light on the process by which these two categories emerged as different, and also shows how opportunities for resistance opened up in the resulting architecture of distinction. Third, this chapter reveals how the cultural tools of antidrug warriors could become available to different groups pursuing their own agendas, some of which posed challenges to the accepted logic of the twentieth century's drug wars. But such opportunities for resistance were limited in subtle and powerful ways. Because drug discourses were so racially charged, this instance of cultural thievery also illuminates how and why embracing a white and middle-class identity could be such a strong, but subtle, temptation for those activists able to lay claim to it.

The Return of the Medical Addict

Before we can understand the Valium affair, we need to know how prescription medicines became a legitimate target of one of America's periodic drug panics. This was in some respects an unlikely development. As we saw in chapter 3, antidrug campaigns had historically targeted marginal groups, and nowhere more directly than in popular culture. Where medical and legal rhetoric had been restrained or at least carefully coded, popular campaigns openly demonized pushers and dope fiends. Rallying public fears helped mobilize state resources to control suspect populations like immigrants, nonwhites, or urban underclasses. These campaigns also helped dramatize cultural distinctions between social groups. Indeed, one of the most common tropes of popular addiction narratives was innocent respectables—young, white, often female—tricked into drug slavery by evil pushers and ineluctably drawn into a world of crime and prostitution. The forbidden pleasures of (often sexual) contact across boundaries of class and race were both sensational and disciplinary, drawing in audiences while educating them about social distinctions.[10]

Prescription medicines, which involved no illicit contact with dangerous social groups, attracted relatively little attention as this cultural system was established. For the first few decades of the twentieth century, popular culture mirrored medical and legal authority in constructing racially charged divisions between medicines and dope. So how did Valium and white middle-class housewives come to fall on the wrong side of the divide?

The answer begins with the contested and ultimately limited effort to rein in barbiturates, amphetamines, and other prescription medicines recounted in chapter 3. As we saw, the drug-law reformers who worked to regulate these medicines were not trying to challenge the fundamental bases of America's antidrug regime. But the process generated new kinds of authoritative, expert information about addiction and about medicines—information, as we saw, that tended to unsettle accepted drug-war narratives. As part of the political maneuvering over proposed new laws, drug researchers, politicians, and others often circulated news about the addictiveness of prescription medicines to the popular media. The spread of such stories was only expanded, ironically, by drug companies' efforts to keep drug regulation a state-level matter. This meant a state-by-state political battle in which at least some powerful groups had incentive to portray barbiturates and amphetamines in a negative light.

As a result, at least some popular accounts of barbiturate addiction began to appear by the 1940s, when, in the absence of federal rules requiring a physician's prescription, many state legislatures began their own regulatory campaigns. These early stories tended to focus on dramatic episodes like a 1945 tragedy in Waco, Texas, in which widespread barbiturate abuse had extended as far as a pair of toddlers drugged by their own pill-addled father.[11] Even after the Humphrey-Durham Amendment in 1951 restricted all barbiturates and amphetamines to a doctor's prescription, addiction experts like Harris Isbell and the congressional hearings in 1955 kept the story of prescription drug abuse before the public eye.[12]

In the 1950s and especially the 1960s a range of different groups, each following their own agenda, found it useful to popularize these newly circulating perspectives about drugs and medicines. And unlike the antidrug stalwarts chronicled in chapter 3, many of them did so in ways that openly drew parallels between the worlds of licit and illicit drugs. Congressional hearings and investigative journalists exposed bare-knuckled drug marketing that seemed to encourage overprescribing. The patients'-rights movement taught that physicians and patients did not always share the same agenda. And in an unlikely (and unintended) alliance, three very different groups—epidemiologists, counterculture youth, and a reenergized medical specialty of addiction treatment—all portrayed ordinary Americans as eager drug takers. Taken together, these new narratives challenged drug-war verities and laid the groundwork for the Valium panic.

The first of these dynamics to appear was a heightened public concern over the commercial success of medicines in the postwar era, a concern powered by congressional investigations and investigative journalism. After World War II, recently consolidated drug houses had basked in the glow of new wonder drugs like antibiotics and vaccines, earning unprecedented profits while enjoying a reputation as trustworthy servants of the selfless medical profession.[13] Increasing competition and ever more intense advertising transformed the character of the once-staid prescription drug business, however, leading to a series of price-fixing and collusion scandals in the 1950s. Tranquilizers drew attention as early as 1958, when the House of Representatives convened hearings on the advertising of Miltown and its competitors. More notably, in 1959 Senator Estes Kefauver, already famed for his televised investigation of organized crime, inaugurated what would be a nearly perpetual series of congressional inquiries into the pharmaceutical industry.[14] Neither Kefauver nor his successors succeeded in curbing pharmaceutical profits, but they did focus media attention on the commercial side of the drug business. "Wonder-Drug Makers Get Handsome Profits from Their Captive Consumers," proclaimed *Life* magazine in one of many reports covering Kefauver's hearings.[15] "Laymen" were "shocked," *Business Week* observed in commenting on media coverage of the hearings, to discover that "executives of the drug houses have the same objectives as other businessmen: to make money."[16]

Kefauver's hearings also attracted the attention of the era's investigative journalists, latter-day muckrakers who used federal research as the basis for an even more damning onslaught of exposés in popular magazines and books. The pharmaceutical industry, with its altruistic rhetoric and aggressive market practices, made an attractive symbol for corporate America gone awry. *Washington Post* reporter Morton Mintz's book *The Therapeutic Nightmare* (1965) established the genre, charging that drug houses manufactured dangerous drugs, stymied efforts to prove or even study those dangers, hyper-marketed—often fraudulently—to achieve maximum sales, and charged exorbitant prices to cover marketing costs while still garnering immense profits. (Mintz's full subtitle: *A Report on the Roles of the United States Food and Drug Administration, the American Medical Association, Pharmaceutical Manufacturers, and Others in Connection with the Irrational and Massive Use of Prescription Drugs that May Be Worthless, Injurious, or Even Lethal.*) Many other such exposés followed, including a flurry of articles and books in the early 1970s and a public investigation by the National Council of Churches.[17] Consumer advocate Ralph Nader joined the campaign, establishing the Health Research Group with physician Sidney Wolfe to issue a stream of press releases and witheringly critical books.[18]

Journalists and government critics were responding to ever more competitive and creative marketing campaigns that increasingly blurred lines, not just between medicines and consumer goods but between medicines and illicit drugs. A series of ads for the stimulant Ritalin in the early 1970s, for example, touted the drug as the answer to noisy urban life, "bad news," pollution, the energy crisis, and even traffic jams ("Who hasn't been bogged down in traffic jams with honking horns, exhaust fumes, and experienced that 'I give up' feeling of resignation?").[19] Valium, meanwhile, was recommended for "marital tensions," the antidepressant Triavil for "empty nest syndrome," and the antidepressant Aventyl for "behavioral drift"—an invented syndrome that, to congressional critics, seemed designed to turn normal mood variability into an illness. Serentil, an antipsychotic, was promoted in 1970 for "the anxiety that comes from not fitting in."[20] These hard-sell campaigns seemed to undermine the special status of pharmaceutical companies as "merchants of health," a noble wing of America's medical system. As one psychiatrist asked at a 1971 congressional hearing, "In the midst of a [drug abuse] pandemic how can we morally and ethically justify the expenditure of one billion dollars per year to promote drugs?"[21]

Adding urgency to such criticisms was a series of local and national studies on drug use that revealed the full extent of America's appetite for sedatives and stimulants. The first high-profile studies, undertaken by epidemiologists from the Psychopharmacology Research Branch of the National Institute of Mental Health, came out in the late 1960s. The studies were careful and thoroughly unsensational. Their authors steadfastly maintained that the medicines were underprescribed in compari-

Ritalin's advertisers went further than most in advocating drug use for everyday problems. *Source:* Ritalin ad, *JAMA*, February 8, 1971. Courtesy of the New York Academy of Medicine Library.

son with the measured levels of anxiety and depression in American society. Nonetheless, their bottom line could not have been clearer: use of prescription mood drugs was shockingly widespread. Half of all adults had tried one in their lives, and almost 20 percent currently used one "frequently."[22] Moreover, the studies seemed to indicate that the numbers were still rising, at least for new drugs like the tranquilizers. To critics this could only be explained by drug company advertising, not growing medical need. One former drug salesman to the National Council of Churches put it this way: "The biggest dope dealer in your community today may well be the good old family doctor, and the pusher supplying him is the tranquilizer manufacturer."[23]

As the salesman's statement suggests, public skepticism was not restricted to the pharmaceutical industry. Physicians and professional medicine too came under heightened scrutiny with the emergence of the 1970s patients'-rights movement. Patients'-rights advocates pushed to enhance access to health care among underserved groups such as the poor and racial minorities. They also sought to address the power imbalance between physician and patient, under the assumption that their interests might not always be the same. Thus they fought to strengthen requirements for informed consent; establish rights for patients to refuse treatment, see their own medical records, and participate in therapeutic decisions; and ensure that patients received due process before involuntary commitment to a mental institution. In response, courts asserted that physicians had an "affirmative duty" to share information fully with their patients, including any risks associated with a treatment.[24]

An important new social movement, second-wave feminism, played a crucial role in advocating for patients' rights. The 1960s and 1970s saw a breathtaking diversity of women's activism on a wide range of issues from welfare rights to race. Some branches of the movement emerged from relatively privileged white contexts and focused on legal and political equality, exemplified by the 1966 founding of the National Organization for Women. Others were forged by younger, more radical women in the civil rights movement; some aimed to reimagine gender altogether; some fashioned a feminism designed to address issues faced by African-American, Latina, or Native American women. Many more focused on particular issues that seemed to have special relevance for (or impact on) women, such as welfare rights, domestic abuse, sexuality, and more. A key vehicle for spreading feminist ideas on these and other topics was "consciousness raising," a deceptively simple process in which women came together without men to talk about their lives. In doing so they often found that what had seemed to be personal problems—the unequal distribution of housework, for example, or unequal work and authority in activist groups—were in fact widely shared and deeply political. Consciousness raising could be local, ad hoc, and informal, and it had became widespread by the early 1970s.[25]

One of the issues that gave rise to its own dedicated segment of activists was women's health. The feminist health movement began in the late 1960s with local groups dedicated to redressing the power imbalance between women and the medical system. They called for—and often helped build—alternatives to traditional care relationships. The Boston Women's Health Book Collective, for example, was founded in 1968 to demystify physicians' expert knowledge so that women could advocate for themselves in a medical system still dominated by male physicians and crude gender stereotypes. They accomplished this by holding workshops, delivering lectures, and publishing what would become the landmark women's health book *Our Bodies, Ourselves* (1971). Another founding feminist health group, Chicago's "Jane," addressed a medical issue that mainstream medicine would not: abortion. "Jane" began providing confidential referrals for (still illegal) abortions in 1969, but members of the group soon educated themselves on the abortion procedure and offered low-cost or even free abortions themselves.[26]

In addition to these locally founded groups of activists, the feminist health movement had its own crusading journalists like Barbara Seaman and Belita Cowen who exposed the dangers of medical therapies widely prescribed for women. Seaman's *The Doctor's Case against the Pill* (1969), for example, raised public awareness of the risks of the birth control pill (at the levels of estrogen used then, it could cause depression, blood clots, heart attack, and stroke). The book generated enough political heat that Congress convened hearings, and eventually passed a law requiring prescriptions to include a "package insert" with complete drug information. The same year, Cowen mounted a one-woman campaign in popular magazines warning women about the "morning after" pill diethylstilbestrol (DES), which she found was neither as effective nor as safe as the medical literature claimed. In 1974, Seaman and Cowen joined other activists in founding the National Women's Health Network as a vehicle for continued research and advocacy.[27]

An unexpected set of allies helped draw public attention to these developments and connect them into a coherent critique of prescription drugs. One does not usually think of the drug-suffused counterculture youth of the 1960s as antidrug crusaders, and they were not. But they had no special love for prescription medicines, backed by corporations and seemingly devised to enforce conformity. And some highly visible young people delighted in drawing parallels between their drug use and their parents' mood-modifying habits. In San Francisco's famous Haight-Ashbury neighborhood, for example, a drug-scene hangout called the Drogstore was decorated as a conventional pharmacy, mocking mainstream culture while also laying an absurdist claim upon it.[28] The Rolling Stones drove home a similar point with their 1966 hit song "Mother's Little Helper," which bitingly mocked a housewife desperate for her doctor's "little yellow pills" to get her through the day. As one prominent new-generation drug scholar noted, "Every time these supposedly law-abiding, drug-hating adults socialize with each other, every time they relate to other human beings, and every time they work or play at being happy or having fun they use drugs."[29] This was nothing new, of course, but rarely had the ranks of illicit drug users included enough culturally empowered members to defend themselves by pointing it out.

The final ingredient in establishing the preconditions for a prescription drug addiction scare was the emergence of a new kind of drug expert who offered an authoritative explanation for the phenomenon. These experts came from the fledgling medical specialty of addiction treatment, an emerging profession that had its own reasons to acknowledge and even trumpet middle-class addiction to prescription medicines. Beyond their ethical drive to address what they saw as a serious problem, they could only benefit from new patient populations from the respectable classes.

Addiction treatment had continued to limp along during the classic era of narcotics control, but in the new atmosphere of the 1960s it saw a renaissance. Nearly $500 million in new federal funding and rapidly expanding coverage by private insurers helped give rise to thousands of addiction treatment centers and programs. In the early 1970s the field was crowned by the creation of two federal research centers, the National Institute on Alcohol Abuse and Alcoholism and the National Institute on Drug Abuse. Addiction treatment soon became a common theme in popular literature and film, and bookstores' self-help sections bulged with literature, magazines, newspapers, and other elements of what one scholar has called an "alcohol and drug abuse industrial complex." Rising legitimacy was matched by increased cultural visibility, as a parade of celebrities and public figures helped create "rehab chic" by checking in to addiction treatment. Addiction specialists were well situated to persuade mass-media journalists that they, not psychopharmacologists, were the true experts about mood medicines.[30]

By the late 1960s and 1970s, then, the intersecting efforts of a number of very different groups had combined to introduce new ways of thinking about prescription medicines. In one way or another, and for their own reasons, politicians, journalists, patients'-rights activists, counterculture youth, and addiction treatment specialists had all helped challenge the carefully constructed boundaries dividing legal medicines and illicit dope. And their critiques gained a wide legitimacy. Respected epidemiologists looked for—and found—links between parents' use of prescription drugs and children's use of alcohol and other drugs, emphasizing the continuity of behavior between taking medicines and doing drugs.[31] Even noted antidrug warrior President Richard Nixon adopted the new rhetoric when speaking to the American Medical Association:

> We used to say [that drug abuse] is a ghetto problem or it is a black problem . . . But today it has moved from the ghetto to the suburbs, from the poor to the upper middle class . . . We have created in America a culture of drugs. We have produced an environment in which people come naturally to expect that they can take a pill for every problem—they can find satisfaction and health and happiness in a handful of tablets or a few grains of powder.[32]

In 1972, the Consumers Union (publisher of *Consumer Reports*) issued one of its respected product-rating references, this one devoted to drugs. While not sharing the political passion of the counterculture's defenders, this book had the same fundamental premise that all drugs, legal and illegal, could and should be measured by the same criteria, even if the results challenged traditional assumptions. The title said it all: *Licit and Illicit Drugs: The Consumers Union Report on Narcotics, Stimulants, Depressants, Inhalants, Hallucinogens, and Marijuana—Including Caffeine, Nicotine, and Alcohol*.[33] Given the polemical history of drugs in America, the appearance (and popularity) of this even-handed, calmly written mainstream book offered striking evidence of the cultural change under way.

"She Could Even Be You": Prescription Drug Scares in the 1960s

These new attitudes toward prescription drugs stirred a flurry of mainstream media attention to medical addiction. Like the experts and politicians examined in chapter 3, journalists first focused on barbiturates and amphetamines. Beginning as early as the 1940s and building toward a crescendo in the 1960s and early 1970s, this early prescription drug scare raised unexpected challenges to conventional antidrug wisdom even while remaining entangled in drug-war politics. Although they did not address Valium or the tranquilizers (that would come later), their scrutiny of white-collar medical addiction paved the way for the Valium panic and is worth examining in some detail.

Stories about barbiturates and amphetamines bore many of the hallmarks of a classic drug panic. A 1967 article in *Look* magazine, for example, informed readers that drug abuse had escaped from its traditional home in the ghetto and invaded the medicine cabinets of middle America. An old and shrinking population of "urban, poor, colored" narcotics addicts, the article maintained, had recently given way to "millions" of "white and affluent" Americans who "can't sleep, wake up or feel comfortable without drugs." Drug problems had escaped the slums and were now sweeping through "white America: Junction City, Kans.; Pagedale, Mo.; Woodford, Va.; Plymouth, Mich.—places with apple-pie smells and wind-snapped flags."[34]

The late 1960s was a propitious moment for a new racially inflected drug "invasion." Headlines were already attuned to the more fundamental challenges of racial integration, and addiction easily took its place as the latest wave of "inner-city" encroachments on "normal" America, jumbled in with school-busing, African Americans moving to the suburbs, and interracial marriage as depicted in such movies as *Guess Who's Coming to Dinner* (1967). These racial fears were interlaced with other middle-class anxieties—about changing social values as well as an economic and cultural status threatened by the end of postwar boom years.[35]

Tales of addiction among respectables had always played upon just such racial, economic, and cultural fears, and this prescription drug panic was no exception. *Business Week*, for example, warned in 1970 that "if you think the 'drug culture' is pretty much restricted to the campus or the ghetto, you might find some surprises by taking a look around your office—and maybe even at yourself." Why? Because "some of the same drugs peddled illegally in the street can find their way into your bloodstream, too—albeit under more respectable labels and usually through a doctor's prescription."[36] In 1967 the *New York Times* opened a five-part series by announcing, "A Growing Number of America's Elite Are Quietly 'Turning On.'"[37] A respected drug epidemiologist, noting that women were the primary users of barbiturates and amphetamines, told *Ladies Home Journal* readers in 1971 that "Times Square prostitutes" and other fringe elements were not the nation's only or even most numerous female drug takers. Instead, "the typical woman who uses drugs to cope with life is an average, middle-class American—one of the folks next door. She could even be you."[38]

The shadowy street and ghetto addicts hovering at the edges of these reports highlighted the cultural dissonance of imagining drug abuse as a central facet of respectable middle-class life. The "invaded" were essentially different from the "invaders," and the sensationalism of the stories came from their linkage through the suspect pleasures of drugs. That the world of "Times Square prostitutes" might be hiding behind the medicine cabinets of the "folks next door" was at once titillating and deeply disturbing. Proper middle-class life itself might put one at risk of addiction! As *Redbook* noted, these were the medicines to which "our families, our friends, and we ourselves are most exposed."[39] The typical pill user, the *New York Times* explained, might even have begun on a quest for therapy—for "support" rather than "fun," seeking "not to stand out from the crowd, but to blend into one and to function the way he thought everyone else did."[40]

The story of "David M.," recounted by the *Wall Street Journal* in 1969, was an archetypal example. A "breezy young salesman" with a fear of flying, David discovered drugs when his physician prescribed a sedative before one of his frequent plane trips. "It worked so well," the *Journal* reported, "that David, prescription in hand, soon began trying other mood-changing drugs when he felt the need." Amphetamines helped him get "up" for business meetings, barbiturates soothed him at the end of the day for sleep. Discussing his daily regimen of half a dozen pills, David exulted, "I never have to feel tired or depressed any more; I don't know how I ever got along without pills."[41]

Other stock characters included pill-popping housewives like "Betty Ann," one of three women profiled in *Ladies Home Journal* in 1971. She and her husband "Don," an ambitious mechanic, lived in a "pleasant, middle-class suburb in a city in the southwest." To raise money to buy a house, Betty Ann got a job selling cosmetics door to door. Nervous about her looks, she was relieved when a physician prescribed amphetamines to help her diet. She lost the weight but never stopped taking the drugs. As she told the *Journal*, "Don likes me thin. And *I* like me better this way, too. I feel pretty, and I'm not afraid of people the way I used to be."

Slowly, her intake increased to thirty pills (350 milligrams) of Dexedrine daily. "On the drug scene," the author somberly informed his readers, "daily ingestion of 75 or 100 milligrams of amphetamines is classified as advanced drug abuse. Betty Ann, age 28, is a speed freak."[42]

Most of these stories focused on what was portrayed as ordinary, even archetypical use of the medicines by run-of-the-mill Americans. They featured no contact with shady characters of any sort. Indeed, people like "Betty Ann" and "David" were defined in part by their resolutely respectable middle-class lives. The narratives thus ventured onto new ground for drug scares by suggesting that something might be wrong with middle-class culture itself. *Redbook*, for example, debunked the notion that "accidental middle class addicts" were fundamentally different from illicit-drug addicts: "In reality, the line between them is hard to draw . . . most addicts are not, as myth would have it, perfectly normal people at the mercy of irresistible drugs."[43] *Look* summarized it bluntly: "The trouble lies not in the pills, but in the people."[44] Many articles emphasized that pill-poppers, not their physicians, were the problem, "talk[ing] doctors into writing prescriptions," forging prescriptions, or even stealing to maintain their habit.[45]

Perhaps the most intriguing counterpoint to the invasion narrative was a genuinely new cultural space opened up by these stories: what one writer for *Atlantic Monthly* described as the "white-collar drug scene." "Sometimes pill-takers meet other pill-takers, and an odd thing happens," he explained. "Instead of using the drugs to cope with the world, they begin to use their time to take drugs. Taking drugs becomes *something to do*." He then described a "pill party" at which the sophisticated but doped-out guests obsessively swallowed and discussed their drugs while passing around the "Book"—the *Physician's Desk Reference*—and reading its comically staid medical descriptions of each drug's uses.[46] The *New York Times* similarly described affluent pill-swallowers as belonging to their own "drug scene" just like "the unemployed Puerto Rican mainlining heroin in a rat-infested Harlem tenement and the barefoot hippie taking LSD in a Haight-Ashbury pad."[47] This notion of a white and affluent prescription-drug culture also made its way into the era's popular literature, for example in Joan Didion's stylish novel *Play It As It Lays* (1970) and Jacquelyn Susann's *Valley of the Dolls* (1966), which was made into a movie in 1967. In these classic period pieces, barbiturates were a ubiquitous and unquestioned option for wealthy, fashionable, but miserable women protagonists.[48]

However unhappy or unpleasant, these drug scenes were far removed from the hothouses of deviance, violence, and perverse interracial sex featured prominently in drug stories from the classic era of narcotics control. If their denizens were not exactly pillars of middle-class morality, neither did they appear to be grossly abnormal. Indeed, what made them objects of fascination was their seeming typicality, suggesting not individual aberration but a class-wide failure of values. Journalistic descriptions of them opened a long-ignored cultural space in the saga of addiction, one that dramatized troubles entirely internal to the nation's comfortable classes rather than posed from without by fiendish pushers. It still emphasized a seemingly natural split between medical and street addicts but did so in a way that brought out parallels as well as contrasts between the two. Against the backdrop of American drug wars, "She could even be you" was a radical slogan.

By the mid-1970s, then, new challenges had arisen to traditional antidrug politics, including the central premise that drugs and drug users were fundamentally different from medicines and medicine users. Some of these challenges were more radical than others, and most continued to traffic in drug-war racial stereotypes, but taken together they represented a noteworthy reconceptualization of American drug politics. As we shall see, when it emerged a few years later the Valium panic would build on both aspects of this inconclusive shift: its campaigners would take the new logic to a more politically radical level while at the same time reaffirming more traditional antidrug tropes.

The Valium Panic

Throughout the prescription-drug addiction scare of the early 1970s, Valium and other minor tranquilizers received virtually no mention. Roche Pharmaceutical's vigorous lobbying staved off "scheduling" for its flagship drug Valium until 1975, and ubiquitous advertising helped maintain the minor tranquilizers' reputation in both medical and popular circles as safe, effective, and useful. Warnings in the FDA-approved texts of advertisements and the *Physician's Desk Reference (PDR)*, for example, continued to emphasize only the risks of prescribing for "dependence-prone" individuals, and conventional medical authorities still viewed the drug's addictive dangers skeptically. The most widely cited medical reference on Valium was a 1973 review in *JAMA* that praised it precisely because "tolerance, abuse, and abstinence are very rare."[49]

Use of Valium thus flourished in the otherwise unpromising climate of the 1970s. Even as national surveys of drug utilization revealed declining use of barbiturates, amphetamines, and prescription painkillers, Valium use seemed to be thriving. Indeed, more than half of Valium prescriptions were for refills that did not require new authorization from a physician—a considerably higher proportion than any other drug in the United States.[50] By the early 1970s, Valium had become the single most prescribed brand of medicine in the nation, with nearly ninety million bottles dispensed yearly.[51] According to the surveys, 15 percent of all Americans had used the tranquilizer or one of its cousin drugs in the past year, 5 percent of them "regularly" (daily for months or more at a time). The numbers were even higher for women, 20 percent of whom reported use in the past year, 10 percent regularly—twice the rate of men's use, and more than could be accounted for by women's greater usage of the medical system generally.[52]

Valium's image began to tarnish after the mid-1970s, however, perhaps in part because it had become so conspicuous; Roche Pharmaceutical's very success in protecting and marketing the medicine began to work against it. In 1975, two political events singled Valium out for public scrutiny. First, after a ten-year legal battle, Roche finally reached an agreement with the Drug Enforcement Administration, which enrolled Valium on the Schedule of Controlled Substances—an event long planned (it coincided with the end of Roche's patent on the drug) but still noticed by journalists already tuned in to prescription drug news.[53] That same year, the recently created Drug Abuse Warning Network, a federal statistics-gathering organization, identified Valium as the single most common drug discovered in overdose victims seen in the nation's emergency rooms.[54]

Paying careful attention to these developments were members of a new medical subspecialty focusing on addiction in women. Led by pioneers like Marie Nyswander, they brought to addiction treatment the logic of the women's health movement. They were thus perfectly positioned to seize upon an epidemic of addiction to "women's drugs" like Valium. In 1975, for example, Nyswander warned readers of *Vogue* magazine that Valium dependency was "a far worse addiction than heroin, morphine, or meperidine (Demerol)," and a greater threat than those other drugs because of its widespread use, particularly among women. In the article, titled "Valium—The Pill You Love Can Turn On You," she explained that "today, probably it would be very hard to find any group of middle class women in which some aren't regularly on Valium."[55]

By the late 1970s, scores of alarmist articles in dozens of popular magazines spread word of the Valium menace. Television's most respected news program, *60 Minutes,* ran a feature on it, and Ann Landers educated readers about it in her nationally syndicated advice column. Three congressional investigations explored the issue, most visibly Senator Edward Kennedy's 1979 hearings devoted entirely to "use and misuse" of Valium and Librium. Along with journalists' reports came book-length exposés, one—*Stopping Valium*—from Ralph Nader's Health Research Group.[56]

Most important in shaping this media sensation were feminists seeking to highlight middle-class women's problems. Their ranks were drawn from many different parts of second-wave feminism, including women's addiction specialists, the women's health movement, and ordinary women who met to discuss Valium in neighborhood consciousness-raising groups. These feminists were well prepared to seize on the possibility of Valium addiction in large part because they had been criticizing tranquilizers and sedatives for years as both symbol and substance of the constraints placed on women of the comfortable classes. As noted in chapter 2, Betty Friedan had inaugurated such reasoning in *The Feminine Mystique*, one of the founding texts of middle-class feminism.

The women's health movement of the early 1970s gave the notion of tranquilizers as an agent of social control deeper articulation and broader expression. Doctors, activists claimed, had always seen women as naturally sickly or as frivolous complainers who took up precious time because they were bored, lonely, and enjoyed the ability to command physicians' attention. Their complaints were psychogenic—"all in their minds"—and thus perfect candidates for therapy with tranquilizers.[57] It is worth noting that these particular gender stereotypes, like tranquilizer use itself, had always been associated with affluent white women. With psychiatric medicines advocated for pregnancy, child rearing, menstrual discomfort, "empty nest syndrome," and menopause, virtually every living woman was a potential candidate for drug therapy.[58]

Feminists countered this logic by arguing that women's complaints were evidence neither of sickliness nor of psychogenic self-amusement but of genuine political grievances. Housewives turned to drugs not out of emotional weakness or pathology but because pharmaceutical companies had "medicalized" these grievances in order to sell drugs, working hand in hand with a sexist society resistant to women's liberation.[59] Why should women "emotionally readjust" to a demeaning and depressing situation? What was needed was not medical therapy for individuals but political "therapy" for society. As one witness at congressional hearings in 1971 suggested, "Tired mothers might do better in working in the National Organization for Women than in taking antidepressants."[60]

By focusing on Valium, feminists used the consumer culture as a field for political activism. The shift to commercialized medicines had helped transform patients into consumers, easing the difficulty of challenging physicians' authority by strengthening the direct relationship between patients and their medical "products." Moreover, Roche's marketing success helped create and maintain broad public interest in the drugs, making them useful symbols for reaching wide audiences. Thus, for example, in *Ms.* magazine's retrospective look at the 1970s, one of the archetypal examples of gender bias in medicine and advertising was a 1972 ad for Milpath, a combination of Miltown and a stomach soother. The ad featured an image of a woman carved into the shape of a chair, accompanied by this text: "G.I. problems making her a fixture in your office? 'Milpath' can cut down her complaints."[61]

But the most important way feminists were able to get publicity for their arguments in the late 1970s was by linking them to sensationalized stories of Valium addiction. Their efforts took advantage of the cultural fascination with drugs long encouraged by antidrug warriors to reach new audiences. Mass-market women's magazines were not known as feminist pioneers, for example, but it was in their pages—in *Good Housekeeping, Redbook, Vogue, Harper's Bazaar,* and others—that the Valium addiction drama largely unfolded. In language familiar to readers of magazines that regularly ran antidrug narratives, authors warned that

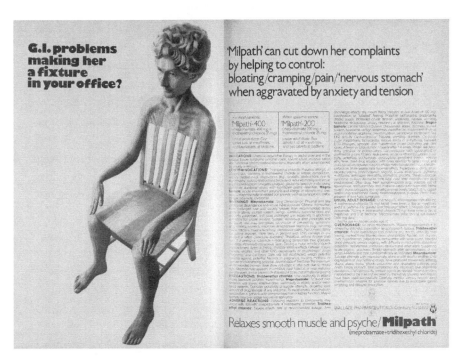

Milpath got unwanted publicity when this advertisement was reprinted in *The Decade of Women: A Ms. History of the Seventies in Words and Pictures* (New York: Putnam, 1980). *Source:* Milpath ad, *JAMA*, February 21, 1972. Courtesy of the New York Academy of Medicine Library.

"nearly two million" or "a staggering number" of "typical middle-class young women" (one expert estimated one in every five) were potential Valium addicts.[62]

The kinds of narratives generated by women's-media journalists were epitomized in the stories of America's two best-known Valium users: former first lady Betty Ford and Emmy-winning television producer Barbara Gordon. While these women could hardly be considered typical, their experiences with Valium became archetypes thanks to media coverage of their stories and the best-selling memoirs both women published in the late 1970s.

Betty Ford had been a dutiful, energetic, and (by her own admission) hard-drinking politician's wife for years when, in the mid-1960s, her physicians prescribed painkillers for a pinched neck nerve. For the next decade she was never without pills. After her husband's brief stint as president, Ford remained a public figure, but friends and reporters noticed a change in the once-vivacious woman. She appeared visibly drunk or sedated in her public appearances, most disastrously during a 1977 visit to Russia, where she was to narrate the *Nutcracker* ballet. She would later recall that visit as a pill-induced "fog." One U.S. journalist noted her "sloe-eyed and sleepy-tongued" performance.[63]

In 1978 Ford called a press conference to announce shocking news: she had been "overmedicating" herself and planned to check in to a naval hospital's rehabilitation program in Long Beach. "It's an insidious thing," she said, "and I mean to rid myself of its damaging effects."[64] Her son, Steve, told reporters that his mother was fighting "a very rough battle against the effects of Valium and alcohol."[65] After eleven days in the program, Ford emerged once more to announce that she was addicted not only to pills but to alcohol as well. The media hailed her announcement as another courageous and forthright move by the woman who had gone public with her battle against breast cancer. She was praised for having confronted the stigma of substance abuse head-on, empowering other women to admit such problems in their own lives.[66]

Barbara Gordon, an award-winning television documentary filmmaker, was a different kind of public figure. Unmarried and childless, she loved the professional freedom she had to pursue her political and artistic passions. Despite this, she described herself in her best-selling memoir as suffering from terrible anxiety, compulsively using Valium to face even mundane activities like shopping or eating in a restaurant. Visiting her psychiatrist to receive the drug was, she later remembered, "routine, like brushing my teeth, a normal part of my life, as it was for most of the people I knew."[67]

Jarred by a friend's death and frightened by her growing reliance on Valium, Gordon decided impulsively to throw her Valium away for good. Her psychiatrist, she later recalled, was angry about her decision and suggested that she switch to a more powerful drug—the antipsychotic Stelazine. When she refused, Gordon wrote, he curtly told her to quit taking Valium all at once rather than slowly reducing her dosage. Gordon did so and experienced psychotic episodes that landed her in a mental hospital twice over the next year. After grueling but enlightening talk therapy in the institution, she finally emerged pill- and anxiety-free, eager to warn other women about the risks of Valium.[68]

Ford's and Gordon's stories as told in the media and by the women themselves highlight the three most important narrative dynamics of the Valium-addiction genre. First, neither woman was portrayed as to blame for her condition. They may have been ignorant of the risks, but they weren't seeking a high and harbored no hidden character flaws that might explain their drug use. Likewise, Valium addicts in the news had always recently quit using the drug, or had only just discovered that dependence might be a problem and had vowed to quit. Indeed, the moment of becoming aware of drug use was, in most stories, identical with the moment of deciding to quit—once the therapeutic rationale for taking Valium had been stripped away, all desire to use it simply disappeared. Thus Ford simultaneously announced her "overmedication" and her plan to rid herself of its "damaging effects." Gordon's suspicions about her drug dependence grew more slowly, but once she accepted her inability to function without Valium, she insisted on going off the drug, despite her psychiatrist's anger and despite the suffering her decision caused her.

Second, women's addiction was attributed to sexist social structures like housewifery and the medical system. Physicians in Valium stories prescribed the drug for genuine grievances like the depression and loneliness of homemaking, the difficulty of adjusting to a husband's constant career moves, and so forth. Or they prescribed it for seemingly random reasons like vaginal infections, skin rashes, or "premenopausal anxiety."[69] Barbara Gordon, for example, portrayed her uncaring and condescending psychiatrist as having practically forced Valium on her, dismissing her fears of dependence with a brusque assurance that "it's not addictive, and you don't have to worry about it." As she told *People* magazine, women should be wary of trusting their physicians: "I was a docile patient," she said acidly, "and ended up in a mental hospital."[70]

Betty Ford's memoirs offered no such criticism of physicians, but other observers were glad to fill in the gap. One addiction specialist told *People* magazine in an interview about Ford that some doctors earn "a great deal of money providing these quick panaceas for their patients. [Others] simply don't know that what they are doing is extremely dangerous, but at this juncture, ignorance of the problem should no longer be an excuse." In a grim irony, she noted, women's more frequent visits to physicians "puts [them] in much greater jeopardy" than men.[71]

Gordon's call for women to stop being "docile patients" and become politically aware advocates for their own interests highlights the third central element of the Valium addiction genre: feminism as a solution to addiction. The transformation from passive patient to assertive participant echoed the consciousness-raising tactic as adapted by the women's health movement, whose members sought to educate themselves so that they could negotiate more effectively with medical authority, or even circumvent it altogether.

Combining this approach with the idea that doctors, not patients, were the ones who wanted women to be taking Valium, many media reports included sections on warning signs of addiction, strategies for recognizing what drug a physician was prescribing, how to question the decision if necessary, and where to look for addiction treatment if the physician resisted. *McCall's*, for example, ran this large-type inset above one of its articles: "Thousands of Americans routinely take these [tranquilizers] to relieve backache or tension—or simply get through the day. When they want to stop, many find they can't. Here is how to recognize the dangers of getting hooked—and where to find help."[72]

These strategies were not only rhetorical. As former Valium addict Cynthia Maginnis told Congress in one of its three late-1970s investigations, women were banding together in their neighborhoods to create new drug rehabilitation programs based on feminist principles. Maginnis herself had helped create one such program, Women-Together, out of her local consciousness-raising group. The women in the group had discovered that they no longer needed to take Valium, Maginnis said, and so they applied for funding as a drug treatment program. The resulting organization aimed not simply to cure addiction but to heal entire individuals through feminist consciousness raising. What they called "outmoded destructive behaviors" would slowly wither away in the "caring, but nonrescuing atmosphere [created] by strong centered feminists."[73]

As Women-Together suggests, feminists of the 1970s had helped frame a strikingly different kind of drug panic. Most antidrug campaigns demonized addicts, pointing to defective character as the cause of drug use and sensationalizing the addicts' threat to "normal" society. The Valium scare embodied the sympathetic, humane, and politically sophisticated approach long favored by progressive drug reformers but rarely seen in practice. Rather than using addiction to air negative stereotypes of drug users, feminists used it to dramatize the hardships faced by essentially decent Valium-using women in a sexist society. Rather than punishment for addicts, they called for greater political awareness and assertiveness. Their success in curbing Valium use through such tactics makes this panic an important but almost entirely overlooked alternative to the punitive antidrug policies studied by scholars and implemented by policymakers.

Of course, all this may look less impressive if no Valium threat ever actually existed. But to say that Valium's critics overstated the danger is not to say that Valium presented no danger at all. Even most skeptics agree that Valium can cause dependence if used regularly for long enough.[74] Regular long-term users may have been a tiny minority of Valium takers, as the drug's defenders pointed out, but given the large numbers involved, even small proportions would have meant a significant problem. A respected national survey in 1979, for example, found that one-fifth of Valium users had been using the drug daily for more than a year. That one-fifth portion represented a full 1.6 percent of all American adults—well over a million potential addicts. Even if only a small fraction of these had a problem, it would still be a serious concern, and the reduced use that followed the Valium scare a welcome relief.[75]

After several years of public criticism, Valium's days as a wonder drug —and as America's most prescribed medicine—came to a decisive end. Use of the drug declined steeply at the end of the 1970s, to under sixty million prescriptions per year—less than half the mid-1970s' peak volume, despite population growth. National opinion surveys suggested that this decline was accompanied by increasingly conservative attitudes about Valium use.[76] Other evidence of Valium's changing fortunes came in Roche's newly defensive advertisements. In front-cover ads in *JAMA* in late 1978, Valium's marketers backed away from their much-criticized hard-sell campaigns by announcing that the drug "*isn't* for common tensions of life—tight schedules, traffic jams, final exams . . . It *isn't* even for many of the tensions and anxieties patients complain about, for example, those due to normal fears and apprehensions about an illness or diagnostic procedure or surgery." Other front-cover ads confronted fears of addiction head-on, advising physicians to "talk over Valium with your anxious patient . . . let them know that you care and that their drug regimen will be limited in duration." An ad appealing to "Feelings vs. Facts" featured the pill speaking in its own defense, claiming itself to be underprescribed with respect to the prevalence of anxiety and insisting on its safety despite "stories [people] have heard about my . . . having the potential to produce physical dependence."[77]

While manufacturers of other tranquilizers quickly stepped in to fill the medical and commercial niche (advertisers for Abbott Laboratory's Tranxene, for example, touted it as a nonaddictive Valium replacement),[78] Valium's downfall reflected feminists' broader success in redefining tranquilizer use itself. The next successful "wonder drug for the mind"—the antidepressant Prozac—would be presented by its champions as nearly the opposite of Valium: it was a stimulant, not a relaxant, and its most famous proponent, psychiatrist Peter Kramer, explicitly deemed it a "feminist drug" for its ability to make women more forceful, less empathic, and more likely to succeed in the business world.[79]

Feminists' groundbreaking strategies in the Valium panic, however, owed at least some of their success to the way they were designed for white middle-class women addicts. The vast majority of popular media stories followed Betty Ford and Barbara Gordon's example in being quite explicit about this, describing corporate wives and professional women suffering horrible withdrawal in "decorator-furnished living rooms" and "ordinary suburbs." The quotation from *FDA Consumer* that began this chapter was a classic example, deploying "junkies" and "street people" for shock value but also to underscore the whiteness and affluence of the Valium addicts.[80]

This dynamic is captured perfectly by Women-Together founder Cynthia Maginnis's testimony before Congress. The story she told followed the classic arc from middle-class housewife to feminist activist. She began taking Valium, she said, when her gynecologist prescribed it to ease her loneliness after relocating for her husband's career. Aided by physicians perfectly willing to prescribe her drugs rather than investigate her troubles, she slowly increased her use until, one day, she overdosed. Shaken, she joined a neighborhood consciousness-raising group, where she developed a sharper political awareness of her situation and worked to achieve self-realization through a range of therapies, hobbies, and other activities. This group became the basis for the Women-Together drug rehabilitation program.[81]

Maginnis described the program as designed for women who were "not the same as people using illicit drugs, who are often court referral cases." It was, she continued, an important alternative to programs "designed to deal with people very different from me." While she did not elaborate on these differences, her reference to "court referral" and her own resolutely middle-class story spoke eloquently on their own. A later witness at the hearing voiced similar concerns, complaining about the "humiliating" red tape of traditional rehabilitation programs and pleading for alternatives to regimens "targeted to the hard core heroin addict, persons who would be more likely to commit violent crime to get funds to maintain the habit."[82]

Like other anti-Valium crusaders, Maginnis found that her whiteness and affluence helped her gain the sympathies of men and women who shared her background. Faced with longstanding cultural traditions of assigning individual moral blame to drug users—and treating women addicts as willful gender deviants—Maginnis turned for support to another culturally available narrative, one that cast white middle-class women as innocent and pure. Her plea for understanding of "people like her" echoed the travails of professional women and white-collar housewives suffering from "the problem that has no name."

One of the central claims of the Valium affair was, in essence, that these white-collar problems, and thus Valium addiction, were fundamentally different from those of the street junkies used as sensationalist decorations in the stories. To accept feminism as a drug policy for Valium did not imply accepting political awakening and activism across the board as a solution to the nation's drug problems. Like other historical instances of what George Lipsitz calls a "possessive investment in whiteness," the terms of the cultural transaction were clear.[83] Anti-Valium campaigners succeeded because they were able to present themselves as an exception; their concerns could be addressed, even in feminist terms, because they raised no challenge to the broader policing agenda of the war against drugs. That this approach was so effective is an instructive lesson in the challenges facing those who would build cross-class and cross-race alliances.

By the 1980s, when the Valium panic (and Valium use) had waned, the political atmosphere surrounding drugs in America had also changed. Few influential voices still blamed social injustice rather than addicts for the nation's drug problems, as resurgent antidrug warriors rallied around simple enunciations of personal responsibility and a renewed focus on inner-city nonwhite drug users. In the revived war against drugs, "She could even be you" became First Lady Nancy Reagan's "Just Say No." If there truly had been a window of opportunity for organizing a broad-based challenge to drug-war politics in the 1970s—a big if—it certainly seemed to have closed a decade later.

Ultimately, the Valium scare deserves attention on two grounds. As an episode in drug history, it offers an alternative template for cultural responses to a perceived drug problem—one that encourages less puni-tive, and possibly more effective, medical and regulatory campaigns. For good reason, scholars have tended to focus on how the state has used antidrug campaigns to police minorities and other stigmatized groups. The Valium affair shows how other groups like feminists could shift the political rhetoric of drug panics to their own agendas, improbably trans-forming the drug war into a kind of civil rights campaign.

As an episode in the history of second-wave feminism, the Valium scare offers a case study in what might be called the limiting successes of the Friedan wing's brand of activism. When they focused on middle-class concerns, it helped them reach culturally privileged women—a powerful political constituency—but did so at the cost of obscuring other kinds of women's issues, in effect making a bid to define feminism itself as con-cerned with the white-collar classes. Given the rich diversity of second-wave feminism, and feminists' often-sought but rarely achieved hopes of building alliances across lines of race and class, this was a success dearly paid for indeed. One can only wonder what the Valium panic, or any other drug scare, would have looked like if powered by a more inclusive effort to define and combat the social underpinnings of addiction.

Conclusion

The Valium scare capped a decade of serious setbacks for those who saw the tranquilizers and other psychiatric drugs as revolutionary, almost utopian medical advances. Some older drugs, like barbiturates and amphetamines, were now strictly regulated. Valium and other more recently developed substances escaped with relatively light restrictions, but attitudes toward tranquilizers had shifted decisively. No longer were popular fears focused on the mind control that might become possible through ever more potent wonder drugs. Instead, the media was rife with stories about minds out of control, addicted to just another much-hyped brand of dope. Valium, the public face of psychopharmacology, had come to emblematize the nation's systemic problems rather than offering a respite from them.[84]

At the end of the 1970s, then, psychopharmacology's cultural power had been seriously weakened.[85] Few voices still called for a drug-induced countercultural revolution, and an intensified cultural emphasis on authenticity seemed to bode ill for drug treatment as a path to a healthy self. But developments in psychopharmacology were poised to change the cultural landscape yet again, as a new generation of antidepressants, led by the celebrity drug Prozac, supplanted Valium in the spotlight. Promoters of Prozac claimed to have found in the drug the answer to the Valium crisis: the new drug was not addictive, and it would return its users (especially women) to authenticity—to themselves—not to "normalcy."

CHAPTER FIVE

Prozac and the Incorporation
of the Brain

ONE OF THE MOST unlikely celebrities of the early 1990s was an anonymous psychiatric patient known only by the pseudonym "Tess." Tess was a rags-to-respectability corporate success whose unhappiness and chronic romantic misfortunes led her to a psychiatrist's office. The psychiatrist happened to be Peter Kramer, who prescribed her the new antidepressant Prozac and described the near-miraculous results in a best-selling book, *Listening to Prozac* (1993). Before medication, Kramer told readers, Tess had been somewhat apologetic as a manager, felt unattractive to men, and was "drawn to tragedy" as a caretaker. On Prozac she was transformed: she became a less conciliatory and more confrontational manager; her romantic life became more "satisfying"; and she lost her martyrish compulsion to "take responsibility for the injured." Kramer concluded that the drug had done something unexpected: instead of returning her to a "premorbid" state—curing her of depression—it had altered her personality. In the book's most widely quoted language, Kramer held that the technologically sophisticated Prozac had made Tess and other patients "better than well," pointing toward a new era in which "cosmetic psychopharmacology" would help people sculpt socially attractive personalities.[1]

Kramer's portrait of a drug that not only cured illness but allowed relatively healthy people to manipulate their identity—the essence of their selves—helped Prozac capture public attention in a way that no medicine had since the heyday of Miltown and Valium. Prozac had already earned moderate fame as a wonder drug since it became available in 1987, but Kramer's book catapulted it to true blockbuster status. Prescriptions skyrocketed as it became one of the most widely used medicines in America and across the world, one of only a handful of drugs earning more than $1 billion annually. Prozac became a cultural celebrity, too: a wash of reports deluged the popular media, Prozac memoirs carved an enduring niche on bestseller lists, and Prozac-inspired jewelry even became briefly fashionable among the Hollywood set. After decades of bitter news, blockbuster drugs were back.

Kramer did not invent the Prozac phenomenon, although he played an important role in packaging and circulating ideas about the drug. As with the blockbusters before it, the Prozac craze emerged from a complex historical interplay of medical, marketing, and cultural dynamics. For two decades several different groups—drug advertisers, middle-class journalists and cultural critics, and a range of medical specialists—had been trying, each in their own way and for their own reasons, to spread awareness of depression as a public health problem. In particular, they sought to rescue depression from its reputation as a serious psychosis and to educate the public about the "milder" versions that they felt were nearly epidemic across the nation. Through their efforts depression slowly took prominence as the main inheritor of the "neurasthenic tradition"—an illness of affluence that served as an arena for purveying and contesting ideas about gender among the nation's white-collar classes. By the 1970s at least some popular and medical observers were ready to declare that the "age of anxiety" had given way to the "age of depression."

Far less successful, however, were efforts to increase the use of antidepressants, whose unpleasant and often dangerous side effects generally restricted them to serious, suicidal depressions. In fact, antidepressant use actually declined in the 1970s as feminists and other drug critics recast psychiatric medicines as addictive agents of conformity and obstacles to authentic self-realization. Prozac revived the dormant notion of blockbuster drugs only after a different set of actors successfully reframed public debate. Responding to fears of addiction and oppression, both biological psychiatrists and drug marketers produced a vision of technologically crafted drugs emerging from newly sophisticated brain sciences. These drugs would permit control of the brain so precise that it would move doctors beyond curing illness to providing a nearly unlimited range of consumer choices for custom-built selves. Medical and commercial promises developed in unison around these drugs, fundamentally linking the project of "cure" with the logic of the consumer culture—and separating both decisively from the stigma of addiction.

After decades of unfulfilled predictions that such miracle medicines would soon be available, advocates found their champion in the late 1980s with the antidepressant Prozac. Prozac was the first genuinely new kind of antidepressant since the 1950s, and it had not been "discovered"; rather, it had been purposefully created in response to theories about how the brain functioned. As a result, supporters claimed, it was selective and subtle enough to end depression and even remake wounded identities without producing addiction or other side effects. And, not least, according to some observers, it was a "feminist" drug that empowered women rather than sedating them. Prozac and other new-generation drugs, in other words, were the anti-Valium. Supporters gained popular attention in part by engaging the public conversation about tranquilizers feminists had stirred up in the 1970s.

This chapter explores the historical roots of the Prozac furor. It begins with a "pre-history" of depression, tracing how the diagnosis came to inherit and in some ways transform the nervous illness tradition in the 1970s. It then turns to the story of antidepressants before Prozac, examining how these relatively obscure drugs came to serve as emissaries of the new brain sciences. Would-be popularizers of the wonder drugs for depression sought to portray them as high-tech tools that could make genuine happiness and self-fulfillment commercially available at the corner drugstore. The last section examines the Prozac phenomenon itself, arguing that the issues raised by the drug—including public debates over biological psychiatry—were shaped as much by the complex history of medicine, marketing, and identity as by any pharmacological qualities it possessed.

The Gospel of Depression

By the turn of the twenty-first century, depression had become so intimately linked to Prozac that the illness might have seemed to some observers to have been practically invented by Eli Lilly. And yet depression had emerged in recognizably modern form decades before the appearance of Prozac. "Although no major poet has stepped forward to provide an announcement," psychiatrist Leslie Farber wrote in *Psychology Today* in 1979, "sometime during the present decade, the Age of Anxiety became the Age of Depression." Many contemporary observers agreed. In medical journals and popular magazines, depression had begun to earn widespread and fearful respect as "the most common form of mental disorder," "the most common clinical disorder of adulthood," "the most prevalent symptom in the U.S.," "far and away the most common emotional illness of mankind," and, by all lights, "virtually epidemic" in America. As strange as it sounds to Prozac-era ears, depression, as *Good Housekeeping* put it, was "the disease of the Seventies."[2]

Depression had not always been seen as so common. Known for centuries as "melancholia" before earning its modern label at the turn of the twentieth century, it had traditionally been understood as a severe mental illness—a dreaded form of insanity.[3] As late as the 1950s, the American Psychiatric Association's *Diagnostic and Statistical Manual* identified seven of the eight varieties of depression as psychoses, marked by such "malignant" symptoms as suicidal thoughts, agitation, delusions, hallucinations, stupor, and so forth.[4] Even in the heyday of Freud, one of the best-known treatments for such depressive psychoses had been electroshock therapy, the fearfulness of which only heightened the perception of severity. This was hardly the kind of diagnosis a physician would turn to except in serious cases. Indeed, labels like "neurasthenia" and "neurotic anxiety" had flourished as means of avoiding the stigma of precisely such mental illnesses. Not surprisingly, relatively few Americans received the diagnosis, and it figured prominently in neither medical nor popular literature.[5]

Milder forms of depression did not go entirely unrecognized. Historian Nicolas Rasmussen has argued, for example, that prominent psychiatrist Abraham Myerson actively sought to raise awareness of mild depressions as early as the 1920s under the label "anhedonia" ("lack of pleasure"). But Myerson's campaign met with relatively limited success. By the 1950s, prevailing psychoanalytic logic still subsumed milder depressions under the supposedly more fundamental category of "neurosis." As we saw in chapter 1, Freudian psychiatry defined neuroses as unhealthy habits of mind developed in response to anxiety. In keeping with this logic, the *Diagnostic and Statistical Manual* described "neurotic depressive reactions" as cases in which "depression and self-depreciation" served to "allay, and hence partially relieve" anxiety.[6] Such depressions were only one among many possible neurotic reactions to anxiety, and certainly not the most important. They received relatively little attention in medical literature and, according to the *National Disease and Therapeutic Index*, accounted for fewer than one in ten diagnoses of neurosis.[7]

The emergence of the modern tranquilizing drugs gave an additional boost to Freud-inspired theories of anxiety's central importance, since it was anxiety that Miltown and Thorazine were presumed to treat with such remarkable success. In fact, as we have seen, the minor tranquilizers were initially seen as effective antidepressants: Carter Product's 1958 guide to using Miltown listed depression as one of anxiety and tension's "related conditions" and cited approximately thirty studies to support its use for the illness. Several early prescription surveys affirmed that many physicians did in fact turn to Miltown, Librium, and Valium to treat nonpsychotic depressions.[8]

In this context, as David Healy has observed, there was little call for an "antidepressant" medicine. Serious depression was a relatively rare form of psychosis, and milder depression already had its wonder drugs in the tranquilizers. True, amphetamine manufacturers Smith, Kline, and French had sought to capitalize on Myerson's "anhedonia" by marketing Benzedrine as an antidepressant beginning in the late 1930s, but their campaigns were quickly overshadowed by the far more widely prescribed tranquilizers in the 1950s.[9]

Thus when Roland Kuhn and Nathan Kline announced the discovery of drugs that seemed to alleviate depression in the late 1950s (see chap. 1), the new medicines did not become famous and immensely profitable blockbusters. They were brand-name drugs but were advertised only in psychiatric journals as treatments for severe, life-threatening depression. Advertisements for Merck Sharp and Dohme's best-selling Elavil, for example, described the drug as a replacement for electroshock therapy, a claim also echoed by Roche for Marplan. Pfizer touted the quick action of Niamid, because "more time increases the threat of self-destruction" (i.e., suicide).[10]

Depression was a marginal illness in the 1950s, but over the next two decades several developments helped give rise to a network of what might be called depression partisans with a much more expansive conception of the illness as more important and more common than others recognized. This reconsideration began with the discovery of antidepressants in the late 1950s, which spurred a number of physicians and researchers—including prominent drug discoverers such as Frank Ayd and Nathan Kline—to take the illness more seriously. They warned about the dangers of depression, especially the risk of suicide, and instructed physicians in how to use the new drugs available to treat it. In an early instance of "listening" to psychotropic medicines, however, their experiences with antidepressants also changed their ideas about mental illness. They became increasingly receptive to arguments such as Abraham Myerson's that even milder depression was a "pivotal affect"—that is, a root cause of illness—rather than a symptom of anxiety. Beginning in the early 1960s, these new depression experts began to spread the word that depression, not anxiety, was actually responsible for America's epidemic of vague psychic ailments and associated physical symptoms.[11]

In his short 1961 book *Recognizing the Depressed Patient*, for example, Ayd claimed that "of all the psychiatric ills to which man is heir, depression occurs with the greatest frequency." Despite its ubiquity, he argued, depression was underestimated by statistics and medical textbooks and usually went unrecognized. Physicians were too focused on severe, suicidal depressions rather than the milder cases that were "among the most common illnesses encountered by the general practitioner." These milder depressions often came in "silent" or "masked" forms, and their sufferers were thus "considered anxiety neurotics and treated accordingly," to no avail. Depression, Ayd wrote, not anxiety, ought to be on a physician's mind when confronted by a patient with anxiety, insomnia, behavior problems, and mysterious physical suffering.[12]

Over the next decades this reassessment of depression slowly gathered adherents and medical respectability. Kline received the prestigious Lasker Award for his work with antidepressants in 1964, telling his audience (and the *New York Times*) that "more human suffering has resulted from depression than from any other single disease."[13] By the late 1960s and early 1970s, depression research had become an alternative of sorts to Freudian theories of anxiety and had begun to attract psychiatrists interested in looking beyond psychoanalytic paradigms. The earliest and best known of these was "cognitive therapy," developed by psychiatrist and former Freudian Aaron T. Beck. This approach called on depressives to recognize their negative, self-critical thought patterns and train themselves to stop such thoughts and replace them with more positive ones. Another major approach was "interpersonal therapy," developed by Myrna Weissman and Gerald Klerman. It held that depression resulted from dysfunctional relationships that could be improved through therapy.[14]

Depression partisans also took their campaign beyond the world of professional medicine, seeking to educate the public as well. These efforts met with mixed success throughout the 1960s, when both the medical and cultural purposes of the nervous illness tradition were being well served by anxiety. Popular accounts only sometimes preached the message of depression's ubiquity and severity as an illness, while others referred lightly to "the blahs" and advised a range of simple diversions (take a drive, etc.) to counteract them.[15]

Efforts to reach the public enjoyed much greater success after 1972, when a U.S. senator's experience with the illness generated national news headlines. Senator Thomas Eagleton was the Democratic candidate for vice president that year but had to resign from the ticket after admitting that he had been hospitalized and given electroshock therapy on three separate occasions in the 1960s. Eagleton's downfall highlighted the stigma still attached to serious depressions. When hospitalized he had publicly admitted only to a "gastric disorder" and a "virus" (this is also what he told the McGovern campaign), and when finally forced to make a public statement in 1972, he initially avoided using the word "depression" and repeatedly referred to himself as having suffered from "fatigue" and "nervous exhaustion"—classic symptoms from the nervous illness tradition, which were not yet fully linked to depression.[16] Electroshock therapy, however, permitted little wiggle room, and Eagleton stepped down, his mental fitness under question.

If Eagleton was unable to escape the stigma of mental illness, however, depression experts were able to use his case to spread their message of awareness, sympathy, and treatment. Eagleton's case put depression into the headlines, and the resulting news coverage helped experts drive home the illness's severity while also providing an opportunity to plead for greater recognition of the vastly more common milder versions. The American Psychiatric Association, for example, convened a panel to field questions from journalists and inform the public that depression was dangerously underrecognized; in fact, it was "the most common form of mental disorder." They also instructed Americans in how to identify the symptoms of such milder, nonsuicidal depressions: sadness, anxiety, fatigue, insomnia, loss of appetite, and decreased or heightened sex drive; trouble with concentration, memory, and decision-making; and "innumerable" vague physical complaints centered around the head, chest, and digestive system.[17]

Eagerly joining the campaign to raise awareness of depression were advertisers for antidepressants. Indeed, one of the reasons Ayd's *Recognizing the Depressed Patient* attracted such notice was that Merck Sharp and Dohme distributed it free of charge to every physician in America. Marketing campaigns were initially restrained, touting antidepressants in psychiatric journals as an alternative to electroshock for serious, life-threatening depressions. They soon turned, however, to the same kinds of strategies that had been so successful for the tranquilizers: promoting their wares for a range of common problems in general-circulation medical journals. With this shift—ironically a return to earlier efforts to market amphetamines in the 1940s—they began to vie more or less explicitly for the same therapeutic territory as the tranquilizers, advocating antidepressants not for a unique or discrete illness but for the vague symptoms and everyday emotional problems characteristic of the nervous illness tradition. Indeed, many early antidepressants were actually tranquilizer-antidepressant combinations.[18]

Carter Products pioneered this strategy with its antidepressant Deprol, a drug whose most important ingredient was meprobamate (Miltown). In 1963, when other antidepressants were still presented in the psychiatric journals as staving off suicide and pushing electroshock into history's dustbin, Carter seeded *JAMA* and other wide-circulation publications with Deprol advertisements aimed at everyday patients and their troubles. According to one such advertisement, Deprol was indicated for family problems, "fear of cancer or other life-threatening disease," and the emotional fallout from arthritis, obesity, hay fever, and cancer, among other things.[19] Needless to say, each of these could be the cause of real psychological suffering, but including them as indications for Deprol certainly expanded the meaning of depression.

It did not take long for other pharmaceutical houses to follow suit. By the mid-1960s Triavil, Aventyl, Elavil, and other major brands had joined Deprol in *JAMA*, and by the end of the decade they had begun to match tranquilizers both in number and in prominence of position on the front and back covers. They also looked like tranquilizer ads, employing the same stock characters, imagery, and creative medical reasoning. Advertisers for the antidepressant Aventyl, for example, provoked regulators' ire by inventing the new illness of "behavioral drift," and advertisers for the stimulant Ritalin launched the "environmental depression" series promising to cure pollution, traffic jams, cultural conflict, and even the Vietnam War (see chap. 4). A 1965 Triavil ad described the drug as "widely useful in everyday practice" and offered as typical patients a troubled and headachey housewife, a nervous and bloated businessman, and a menopausal woman.[20] Four years later the drug's advertisers recommended the drugs for "break-ups": the image was of a smashed teacup, and the text warned that personal or financial break-ups—indeed, "any significant loss or severe blow to self-esteem"—could mean danger. Advertisers for competitor drug Elavil warned in 1965 that "loss of any emotionally valued object can be a significant precipitating factor in depression," so physicians should thus look for evidence of depression when presented by "real or fancied [imagined]" loss of health, appearance, a loved one, material goods, or self-esteem.[21]

Sometimes the effort to tap the tranquilizer market was even more direct. In 1966, advertisers urged physicians to reach for the antidepressant Elavil "instead of the 'sleeping pills' she asks for," because "insomnia is often a sign of depression."[22] In 1971, Triavil's marketers advised physicians to consider the drug "before you prescribe a tranquilizer alone for a patient exhibiting anxiety."[23] Other campaigns invoked—and expanded upon—Frank Ayd's language of "hidden" or "masked" depressions to explain why anxiety and other emotional ills might merit treatment with an antidepressant. "When anxiety accompanies depression," ran a 1969 Elavil ad, "medical opinion frequently regards them as one basic entity." The ad's striking image showed two mobster caricatures joined at the hip, one anxious and one grim-faced.[24] This kind of logic also appeared in a new kind of advertisement advising doctors to look beyond their patients' surface appearance—and their own claims about their health—to find the depression lurking within. As one series of ads for Aventyl sympathized, "Do you have patients who try to hide fear behind bravado?" "Frustration behind conformity?" "Anger behind charm?" "Anguish behind arrogance?"[25]

These advertising campaigns in combination with efforts by depression experts were apparently quite effective in spreading the gospel of depression. As the more broadly defined version of depression became better known, epidemiological surveys began to substantiate predictions of the illness's ubiquity: the same kinds of studies that had once demonstrated anxiety's wide prevalence now found depression in vast swaths of the American population. Diagnoses of nonpsychotic depression began to rise, slowly but steadily; according to the *National Disease and Therapeutic Index*, they reached a rough parity with "anxiety" in the early 1970s.[26] In 1978 a physician writing in *Vogue* magazine captured the changing culture succinctly: "a 'nervous breakdown,'" he instructed, "is generally a depression."[27]

Physicians and advertisers were able to raise awareness of depression so successfully, in part, by associating it with the cultural dynamics of the nervous illness tradition so effectively exploited by anxiety's evangelists in earlier decades. Despite regular avowals that depression spared no class or category of Americans, for example, almost all public discussion of the "new" milder depression took place in contexts of affluence. Studies showed that, like anxiety, these milder forms were diagnosed primarily among the white and affluent, even as symptoms of depression (usually untreated) were far more prevalent among the nonwhite and poor. According to at least one group of researchers, this reflected a difference between the illness of depression and the sadness of people confronted with genuinely difficult circumstances. "Depression," they argued, "must be distinguished from sociocultural conditions—for example, it would be inappropriate to treat the *apparent* depression of the aged rural Southern Negro when this depression is really an adaptive social gesture."[28]

As depression began to eclipse anxiety in public awareness, moreover, advertisers and cultural critics increasingly used the illness as a means to tell remarkably familiar stories about white-collar men and women. Men in antidepressant advertisements, for example, continued to wear business suits and ties, and many expressed their illness through complaints about stomach pain or discomfort.[29] One 1968 ad for the mixed antidepressant-tranquilizer Etrafon followed the "caveman-within" reasoning by quoting Margaret Mead and using the imagery of an aborigine next to a suit-and-briefcase man (see chap. 2).[30] The ads also paralleled tranquilizer logic by portraying depression as a threat to masculinity: depressed men lost their decisive vigor, or had been dislodged from male roles through retirement, loss of a wife, and so forth.[31] When depressed men like Senator Eagleton entered the public spotlight, it became fashionable to mine the ranks of history's great men for depressives: Abraham Lincoln was the favorite, along with Vincent Van Gogh, Winston Churchill, and a parade of other luminaries including several former U.S. presidents. Observers found that depression, just like anxiety, was "an illness of the ablest," disproportionately striking "ambitious, energetic and successful" people like Eagleton.[32] In the first well-known depression memoir, *A Season in Hell* (1975), Percy Knauth described his final recovery from depression in unambiguous terms: "For the first time in all those years, I knew where I stood. It left me with my manhood unassailed, it restored my worth as a man in my own eyes."[33]

But men were far less likely than women to be the subject of attention. Especially by the mid-1970s, depression was considered a woman's illness in both popular and medical culture. Media coverage overwhelmingly appeared in the pages of women's magazines, where case histories tended to be of privileged women—the "kind of woman other women envy"—struck inexplicably by depression. Drug advertisements were far more likely to feature women than men (not necessarily true for tranquilizers, as we have seen). And epidemiologists reported a two-women-for-every-man gender discrepancy in both symptoms and diagnoses.[34] As Nathan Kline told *Vogue* in 1975, he heard "essentially these same words" from hundreds of women each year: "I have everything any woman could want and I can't understand why I don't enjoy or appreciate it." These women, he assured readers, were "not 'ungrateful bitches'" but "victims of a great unrecognized disorder."[35]

As with tranquilizers, advertisements for antidepressants tended to reinforce women's housewifely roles. Many of them, as we saw in chapter 2, portrayed housewives' dissatisfactions as an illness. In 1972, for example, advertisers for Triavil exhorted physicians to check for depression if a patient "finds it 'almost impossible to start housework every morning.'" Sinequan's advertisers featured a woman listing unhappiness with her marriage as a symptom of depression, along with "headaches, diarrhea, this rash on my arm," while Vivactil's marketers promised their drug would "get the patient moving" (down the stairs with the laundry) and then to "get her mood improving" (with her alluring expression inviting a sexual interpretation; see chap. 2).[36] Other advertisements warned that

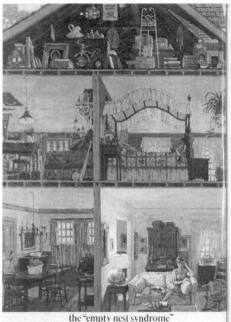

Medicalizing the "empty nest syndrome." *Source:* Triavil ad, *JAMA* January 6, 1975. Courtesy of the New York Academy of Medicine Library.

depression could strike when the homemaker role was undermined by the departure of the men in her life through death or maturation. "For him, commencement," a 1968 advertisement for Tofranil ran. "For his mother, the beginning of his career may seem the end of hers. The end of feeling needed and useful." Triavil's marketers were more succinct, touting their drug for "empty nest syndrome" in 1975.[37]

Gender crusaders also tended to see in women's depression the same kind of crisis they had seen with anxiety. Feminists, for example, followed Betty Friedan's example and argued that women were ill because of the limits society imposed upon them. A good example of such reasoning was *The Book of Hope* (1976), a self-help book for depressed women written by psychiatrist Helen De Rosis and successful women's-media editor and columnist Victoria Pellegrino. *The Book of Hope* interpreted the "chronic, low-grade depressions" of its mostly well-educated and affluent women as a means of coping with the feelings of "rage, grief, and hopeless despair" that came from living according to society's "shoulds" rather than following their own true needs. "For some women, there is dissatisfaction with the housewife role," they argued, "and still there is not yet an equal place for women in the world at large." Echoing Friedan, De Rosis and Pellegrino avowed that "you have a right to be yourself" and urged readers to consider a range of fairly rarefied possibilities for self-expression: skydiving, starting a business, learning Russian or judo, becoming an essayist or poet—"no matter what, we're on your side." Such arguments could also be found in popular magazines like *Good Housekeeping* and *McCall's*. In at least some quarters, the "problem that has no name" had finally found one.[38]

At the same time, however, more "traditionalist" gender critics agreed with drug advertisers, seeing depression as evidence that women were becoming sick for exactly the opposite reason: they had strayed too far from their "natural" roles as wives and mothers. Echoing the logic of emerging antifeminist groups such as Phyllis Schlafly's "Eagle Forum," these critics argued that feminism itself might be the root cause of the depression epidemic among women. As explained by Maggie Scarf, author of the bestseller *Unfinished Business: Pressure Points in the Lives of Women* (1980), women had been programmed by evolution and culture to place enormous importance on interpersonal relationships. Close and nurturing connections to family, friends, and children defined the womanly lot, not the pursuit of money, autonomy, or political power. And when these relationships were unsatisfying or severed, depression struck. The thriving feminist movements of the 1970s, Scarf suggested to *Vogue*, had loosed an epidemic of the illness because under its auspices "the female's traditional roles as caregiver and nurturer are being questioned (and demeaned, to some degree) . . . women learn early that they should be ashamed of the very set of qualities which are particularly theirs." The implications of her findings were stark: "Women's struggle to be autonomous probably flies not only in the face of powerful conditioning started in earliest infancy but of biological tendencies as well. For women, the way in which we now lead our lives is, I believe, profoundly antibiological."[39]

Scarf, who described herself to *Newsweek* as "freaked out" by her "sexist" conclusions, relied on the medical reasoning of one of depression's new experts, co-founder of "interpersonal therapy" Myrna Weissman.

For her influential work *The Depressed Woman* (1975), Weissman had studied two groups of women, one depressed, one nondepressed. Differences between the two groups, she reasoned, could be ascribed to the illness. Her conclusions, like Scarf's, echoed those of 1950s-era gender traditionalists: "normal" (nondepressed) women had a "low level of social interaction with the wider community" outside their family and yet "experienced housework as satisfying, parent-child relations as warm and involved, marriages as affectionate and harmonious." Their lives were centered around "a close-knit extended family, relations with whom were characterized by compliance and dependency." Contrary to "other studies that have reported a high degree of pathology in normal subjects," this arrangement produced no "dissatisfactions, friction, or conflicted loyalties." Depressed women, on the other hand, were hostile toward their families. They were physically uninterested in their spouses yet sexually manipulative, "overconcerned, helpless, guilty, and sometimes overtly hostile" as parents. "Interpersonal therapy" had as its goal healing these relationships, not necessarily in ways that addressed power imbalances within them. For example, it might mean helping a woman adjust more comfortably to relocations forced by her husband's job.[40]

Overall, then, depression entered American medical practice and popular culture as a proposed successor to anxiety in the nervous illness paradigm. In medicine, it provided an arena where psychopharmacologists and other researchers with weak connections to the Freudian establishment could pursue new theories without directly tangling with psychoanalytic orthodoxy. In the broader culture, depression became a vehicle for the same kinds of argument about gender in affluent America that had animated anxiety in the 1950s and 1960s.[41]

Despite all the similarities, it is worth noting one significant difference. While would-be philosophers and gender crusaders had found value in anxiety as a source of authenticity, virtually no one disputed depression's status as a dreaded disease. In both medical and popular literature it was shadowed by references to suicide, psychosis, and electroshock. Unlike anxiety, which could seem far removed from the ravages of schizophrenia despite their presumed underlying connection, depression remained a unified concept, with milder forms seen just as that—milder forms of a serious illness. It may have been common among affluent populations, but very rarely did it inspire proprietary feelings.

Instead, it appeared as a tragic force that blocked creativity and ended promising careers with suicide. Many observers even described it as a sort of archetypal illness, recognizable because it made people become "not themselves." Indeed, the countless stories of picture-perfect women (and some men) inexplicably stricken with gloom paralleled popular tales of drug addiction, with an underworld threat invading comfortable settings and stealing away innocents.[42]

Antidepressants in the Age of Tranquilizers

Perhaps it had only been a matter of time. Amid all the new attention lavished on depression, *Business Week* broke a story on a potential new blockbuster antidepressant that researchers were hailing as "the key to a whole new era" of medicine. The new drug—still in testing—relieved "emotional depression" even when other medicines failed, and did so with virtually no side effects. Because as many as two out of every five patients who come to the typical doctor's office suffered from depression, the authors pointed out, the market for such a pill was "incalculable." Indeed, industry experts predicted that sales of antidepressants on the whole were set to rise as much as half again over the previous year.[43]

This was a familiar story in the era of Prozac—except the year was 1963, and the drug was Eutonyl. And though the article was unusual, it was not the only one. Another article in *Popular Science*, penned in 1962 by Medical and Pharmaceutical Information Bureau employee Lawrence Galton, spoke the gospel of depression ("Ask medical scientists today what the most common of all health problems may be and they'll point to depression") but was also uncharacteristically sanguine about antidepressants. "Drugs are by far the most popular method of treating depressions today," the author wrote, "especially in mild cases." Adding a touch of before-and-after magic to an already overly optimistic story, he described one 24-year-old patient who reportedly "felt like 60" before treatment but "feels like 15" afterward.[44] *McCall's* in 1958 and *Time* in 1965 also prematurely announced magical new pharmacological cures for depression.[45]

This apparent hunger for a wonder-drug antidepressant had little to do with any special qualities of Eutonyl, a copycat drug that never surfaced again in popular media. Nor did it reflect a widespread interest in antidepressants. Indeed, one of the most surprising aspects of depression's newfound visibility in the 1970s was how little change this brought to the reputation of the antidepressants, especially given the importance of psychopharmacologists and drug advertisers in spreading the word about depression in the first place. This is not to say that antidepressants were ignored, but they did not receive the attention heaped on the minor tranquilizers. Despite advertisers' best efforts, they continued to be portrayed in both medical and popular literature, even by their fiercest partisans, as major treatments for serious depression—*not* something for "the common patient in everyday practice," but something that could replace electroshock therapy as Thorazine had supposedly replaced straitjackets. Ayd's *Recognizing the Depressed Patient*, for example, recommended supportive conversation, rest, and perhaps a sedative for patients with mild depression. Antidepressants, meanwhile, were "a satisfactory substitute for shock therapy" for very serious cases involving mania, delusions, or suicidal thoughts.[46] The drugs played a similar role in popular accounts, for example praised by *Vogue* in 1972 as a godsend for "psychotic depressives" (those "wretched human beings who used to fill the mental hospitals, where they had been sent by their physicians or families to wither away") but were left unmentioned when the article turned to "those depressions we call 'normal.' "[47]

The 1970s may have been the age of depression, in other words, but it was not yet an age of antidepressants.[48] Burdened by serious side effects and obscured by the long shadow of the more famous tranquilizers, antidepressants represented only a small fraction of psychotropic drug usage in America.[49] Prescriptions did not even reach parity with antipsychotics until the early 1970s, and by 1980 antidepressants—their star firmly hitched to the increasingly stigmatized tranquilizers—had seen their use decline slightly as they stayed at a respectable (but far from dominating) one-fifth of all psychiatric drugs prescribed (see appendix B).[50] More so than tranquilizers they remained the province of psychiatrists rather than general practitioners, and surveys revealed that they were also more likely to have been prescribed for a formal diagnosis of mental illness.[51] Predictably, relatively few Americans reported having used an antidepressant: approximately 4 percent, claimed one 1972 study of Boston, barely more than antipsychotics. Well over half of those surveyed reported having used an antianxiety drug.[52] When generalists were confronted by patients with psychological symptoms, they usually diagnosed anxiety or mixed anxiety and depression, and overwhelmingly preferred to dole out minor tranquilizers rather than antidepressants, often even for the rare diagnosis of depression itself.[53]

If antidepressants could not match tranquilizers' commercial success or cultural visibility, however, they were considerably more influential within the slowly developing world of biological psychiatry. Miltown and Valium had initially been understood through the Freudian paradigm, but from the very beginning antidepressants served as heralds of new theories of brain neurochemistry and hopeful new visions of a biological approach to mental illness.

The new biological psychiatry had begun with what appeared to be a most productive series of observations in the 1950s. First, researchers noticed that the major tranquilizer reserpine seemed to cause depression or even suicide in some users, suggesting a possibly biological basis to depression. Further study revealed that reserpine depleted the brain of neurotransmitters called "monoamines." Monoamines, like many other neurotransmitters, are bodily hormones that (among other things) act as messengers between nerve cells in the brain. They are released at the end of one cell and connect to "receptors" in the next, triggering it to fire; they are then ejected and either "cleaned up" by janitor-like enzymes ("monoamine oxidase," or MAO) or sucked back in to the original cell. When reserpine caused levels of these monoamines to decrease, it produced a decline in overall nervous activity that seemed to correlate logically with a depressed person's lower energy levels and lack of spark or passion. Seeming to confirm this reasoning were the two classes of antidepressants discovered in the late 1950s, both of which increased the amount of monoamines: "tricyclics," by blocking their reabsorption into the initial cell, and "MAOIs," by inhibiting the janitor enzymes. Depression, it seemed, might be a relatively simple disorder caused by incorrect levels of just a few chemicals in the brain.[54]

One powerful appeal of this model of depression was its suggestion that neurochemistry might be a relatively straightforward system, with mental disorders caused by simple defects. A similar optimism could be found in most influential biological theories of mental illnesses like schizophrenia. In particular, researchers sought to identify the single monoamine that caused a disease, hoping that the disease could then be treated with one of medicine's fabled magic bullets. High levels of dopamine, for example, were widely believed to cause schizophrenia, while low levels of norepinephrine or serotonin were blamed for depression. These simple but influential hypotheses had the salutary effect of defining exactly what caused a disease and establishing clear and measurable procedures for treating it—an especially important factor after 1962, when the Food and Drug Administration (FDA) required drug companies to prove that new medicines were effective for an established disease.[55]

As David Healy and other critics have observed, however, these hopes for chemical simplicity in the brain were rarely supported by evidence. No one has demonstrated that schizophrenic or depressed people have unusually high or low levels of monoamines in their brains. Whatever psychotropic medicines are doing, they are not correcting for any observable imbalance in dopamine, serotonin, or norepinephrine. Even the initial observation that reserpine caused depression turned out to be largely anecdotal and either misleading or outright incorrect (indeed, there was just as much evidence that reserpine might actually act as an antidepressant). In the case of antidepressants, no one has explained why it takes two weeks for the drugs to start working even though they raise monoamine levels immediately, or why the drugs work for only a relatively small portion of depressed patients. Even the most positive studies favored by drug manufacturers show improvement in approximately 60 percent of patients, barely more effective than a placebo.[56]

These and other mysteries do not undermine the importance of biology for mental and emotional illnesses, but they are devastating to simplistic models equating moods with the quantity of a few monoamines in the brain. They suggest, instead, other models for biological psychiatry. For example, the phenomena of the brain might be like the weather: sufficiently complex and self-influencing that complete precision and control are impossible even in theory. Healy, for example, describes a model in which moods are determined by the relationships between neurotransmitters, which themselves may be in constant and very complicated flux. Drugs could hope only to serve as "network management tools" in such an ever-evolving system.[57] Another model emphasizes the difficulty of pinning down causality in a system premised on interactions between genes, biology, and lived experiences. Elizabeth Wilson, for example, has argued that causality can be fully distributed within such a complex system to the point that "no particular element emerges as the originary, predetermining term." Biology, in other words, plays a role in moods and emotions, but we cannot know with certainty that it *causes* moods and emotions, as opposed to developing in tandem with them.[58]

The reason simplistic explanations like the serotonin depletion theory of depression survived, Healy and others argue, is not because they were proven true but because they were so useful for virtually everyone involved: the pharmaceutical industry, which used them to prove efficacy to the FDA and to advertise its wares; professional psychiatry, which used them to claim true understanding of mental illness while highlighting the importance of their prescribing power in comparison with other kinds of therapists; insurance companies, which benefited from cheaper therapy; and patients, who often preferred a physical to a mental illness and were grateful for a simple cure. With such institutional power behind them, it is not surprising that simple models of mood and molecules drowned out more complex arguments that did not promise such easy payoffs.[59]

The simplicity and hopefulness of the monoamine model also helped antidepressants' supporters gain a hearing in the popular media even in the 1960s and 1970s—the era of Freud and tranquilizers. Virtually every mention of antidepressant drugs seemed to include a brief description of how they worked, helping to make antidepressants (unlike tranquilizers) some of the earliest popular vehicles for spreading awareness of the new theories of biological psychiatry. *Business Week*'s 1963 encomium to Eutonyl, for example, explained that the drug worked by blocking MAO: "if too much monoamine oxidaze [*sic*] is present . . . there won't be enough monoamines to carry the nerve signals normally." Lawrence Galton's 1962 article had provided a similar education in *Popular Science*: "It may be that in depression there's too much MAO, too little monoamines. The psychic energizers curb MAO actions, thus making monoamines more available to stimulate the brain." In women's magazines—a key location for discussion of depression and its treatments—such reasoning might be left relatively vague, as when *Good Housekeeping* explained in 1975 that antidepressants "corrected a chemical imbalance in the processes that control the working of the brain," or it could sound confidently technical as when Nathan Kline told *Vogue* the same year that "there is fairly good evidence" that in depressed people's brains "certain 'biogenic amines' are either not produced in sufficient quantity or are destroyed much too rapidly." Monoamine theory also appeared in popular-format books such as Nathan Kline's *From Sad to Glad: Kline on Depression* (1974), Ronald Fieve's *Moodswing: The Third Revolution in Psychiatry* (1975), and novelist Percy Knauth's 1975 memoir *A Season in Hell* ("There is little doubt that I had been suffering from a norepinephrine imbalance," he wrote, "which the antidepressant medication had now set aright").[60]

Despite the best efforts of supporters, however, antidepressants remained relatively invisible during the 1960s and 1970s, especially compared with tranquilizers. Prescribed so much less than tranquilizers, they did not catch the attention of middle-class gender crusaders the way Miltown and Valium had done. Ironically, they also may have suffered from the lack of negative publicity. Even though depression itself was slowly gaining visibility as the latest incarnation of the nervous illness tradition, drug treatment remained curiously unaffected by the cultural debates raging around gender and class politics. It seemed to occur to no one, for example, that use of antidepressants heralded doom for American masculinity, and the drugs generally escaped notice from feminists who saw depression as a consequence of oppression, and also traditionalists who saw depression as a consequence of feminism.

Antidepressants, Consumerism, and the Popularization of Biological Psychiatry

Antidepressants did not, then, ride to cultural fame on a wave of negative publicity. Instead, they became a vehicle for popularizing the new brain sciences and the revolutionary consumer good they would soon provide: the "designer brain." Accompanied by simple explanations of how they worked, antidepressants were a perfect example of the relatively straightforward theories of brain chemistry favored by commercialized medicine. Particularly in the pages of high-end popular media (the kinds that targeted well-educated, affluent readers, such as trade nonfiction books, the *New York Times*, PBS, *Smithsonian* magazine, etc.), antidepressants helped rehabilitate psychopharmacology from the Valium panic by pointing toward the miraculous new treatments soon to be produced by ever more sophisticated brain sciences. New drugs would do more than cure illness (although they would do that). They would enhance, augment, and otherwise improve normal mental and emotional states as desired. As professional psychiatry wrestled with its well-documented shift from "blaming the mother" to "blaming the brain," the broader dynamics of commercial medicine focused popular discussions on the consumerist possibilities of the new technology. Antidepressants, as one of the few examples of the brain sciences in action, became a springboard for replacing narratives of addiction and oppression with those of consumer choice and endless possibility.

An early but important example of how consumerist fantasies could influence biological thinking can be found in Nathan Kline. Interviewed by the *New York Times* in 1957 about his discovery of the antidepressant iproniazid, he mused that the drug's "beneficial action may not be limited to sick individuals" but might "improve ordinary performance . . . of essentially 'normal'" people. In his introduction to Robert de Ropp's mass-market 1957 book *Drugs and the Mind*, Kline was even more optimistic, proclaiming a new "chemopsychiatric era" in which drugs would provide more than mere "surcease from sorrow." "The exaltation of heightened awareness, strong positive affective relationships and the pride of useful accomplishment may, at this moment, be in our grasp," he rhapsodized. (De Ropp's book went on to assert that all human characteristics—"the cruelty of the tyrant, the compassion of the saint, the ardor of lovers, the hatred of foes"—were "chemical processes" that could be produced or halted by drugs or through electrical stimulation of the brain.) Ten years later Kline, speaking at a conference of psychopharmacologists, science fiction authors, and public intellectuals, was prepared to be specific about what humanity could expect from drugs in the future: medicines to prolong childhood; reduce or 'totally circumvent' the need for sleep; regulate sexual response; control affect and aggression (perhaps through lithium in the water supply); prolong or shorten memory; induce or prevent learning; produce or discontinue love; provoke or relieve guilt; foster or terminate mothering behavior; shorten or extend experienced time; create *déjà* and *jamais vus*; and deepen awareness of beauty and sense of awe."[61]

Kline was certainly ahead of his time in imagining the eventual fruits of biological psychiatry, but this strain of thought did become increasingly important as the new brain sciences began to gain visibility in the 1980s. A key step in the development and popularization of these ideas came in the 1970s from researchers studying the effects of opiates on the brain. To affect the brain, they theorized, opiates must mimic some naturally occurring substance in the brain; otherwise there would be no "receptor" for opiate molecules. Following this theory, they discovered the brain's natural version of opiates (enkephalins and endorphins), which turned out to be relatively simple molecules known as peptides. The theories and techniques they pioneered allowed researchers to identify and isolate dozens of new peptide neurotransmitters.[62]

The discoveries were widely hailed as a revolution in brain chemistry that would soon produce an addiction-free painkiller as well as a wealth of other brain-controlling drugs. As *Smithsonian* magazine announced in 1979, "Freud predicted more than 70 years ago that someday 'a special chemism' would be found to explain processes underlying sexual behavior and every mental or emotional event . . . that day now seems imminent. People could one day choose from a wide assortment of morphine analogs tailored to increase creativity and intellectual acuity, or to produce any emotional state—anticipation, reverie, determination, curiosity, calm." An article in *Maclean's* opened by trumpeting the "wonders" that "leading brain researchers" were on the trail of: "Improve your memory! Increase your sexual potency! Relieve your anxiety! Banish your depression! Maximize your powers of concentration! Free yourself from agonizing physical pain!" *Futurist* magazine offered a similar list of upcoming possibilities: a nasal spray to enhance memory; drugs to increase attention span, relieve anxiety and aggression, reduce fear, abolish schizophrenia, and augment creativity.[63]

The chorus of voices grew in the 1980s as a range of researchers and physicians sought to popularize what one reviewer called "the new brain." Psychiatrist Seymour Rosenblatt, for example, teamed with journalist Reynolds Dodson in 1981 to write *Beyond Valium: The Brave New World of Psychochemistry*, which offered a forthright prediction after surveying the current state of psychopharmacology: "We are approaching the age of 'mental face-lifting,' and, like surgeons before us, we will be dealing with the question of whether the methods we have developed should be restricted to sick people." Rosenblatt speculated that drugs might eventually fix basic inequalities between the genders ("Might it not be possible," he asked, "that there will be a groundswell of research to try to undo this anatomical injustice?"), or "sustain the libido, turning the act of coitus into an Olympic marathon." Perhaps most provocatively, he predicted that science was mere years away from "zeroing in on brain centers for human achievement—not just memory but also creativity, musicianship and what makes 'gifted' or extraordinarily coordinated people." Once these were identified, "in a relatively short time the mysteries of virtuosity would be unraveled." How far might *Homo sapiens* go in such self-improvement? "I suspect there's a limit," he wrote, "but we haven't reached it, and right now it's like posing a 'how high is up' question."[64]

Over the next decade Rosenblatt's book would be joined by a flurry of other popular reports from the new brain sciences with titles like *The Brain: The Last Frontier* and *The Amazing Brain*. Some of these were relatively nuanced, such as PBS's multipart series "The Mind," accompanied by an introductory textbook by neuropsychiatrist Richard Restak. These more sober works did not foretell the coming of blockbuster miracle drugs, although they did contribute to the overall story of progress and knowledge that helped make such predictions believable when they appeared elsewhere. Michael Gazzaniga, a professor of psychiatry who wrote popular books on the brain, gives a good example: when discussing the biological nature of love, he reported that a fellow researcher had "actually traced brain circuits that are crucial for triggering this kind of arousal" and concluded that "neuroscience is in hot pursuit of the brain mechanisms of our subtle emotional life."[65]

The more popular a work was, and the later it appeared, the more likely it was to use speculation about future wonder drugs to draw attention to the new brain sciences. The 1987 *Molecules of the Mind* by two-time Pulitzer Prize–winning journalist Jon Franklin is a good case in point. Dramatizing the story of neuropsychiatry as a series of heroic discoveries, Franklin declared that researchers had "converted [psychology] from a form of witchcraft into a laboratory science and ultimately an engineering discipline," putting "the human condition" itself "up for grabs." Given time, he continued, "researchers believe they can develop very specific drugs to reduce appetite, enhance learning . . . even, one top scientist speculates, reduce shyness." He then quoted brain researcher Candace Pert, one of the discoverers of opiate receptors in the brain, imagining a love potion: "I can conceive of a drug, or a battery of drugs, that make some cortical connections ... *zzzt! zzzt!* ... and you will fall passionately in love for life with the face that you're presented with during the next sixty seconds. I can see that." Such a potion was "not wild speculation," Franklin judged, and might help Americans "preserve marriage." On the other hand, "Equally valuable would be a drug capable of breaking the spell of love, a sort of 'antilove potion,' so that divorces can be clean, without bitterness." And, he concluded, "the possibilities go on," including "memory-enhancing drugs" for students, and "intelligence boosters" given "judiciously" to "scientists and diplomats handling international crises." Perhaps most enticing for drug companies, Franklin dangled this possibility: "And—think about this!—each time we discover a molecule we have discovered the potentiality for at least two diseases. Some people will surely have too much of that molecule; others will have too little."[66]

These popularizations of the new brain sciences helped change the tenor of the biological revolution. As we saw in chapter 2, the 1970s saw a great deal of increased scholarly and popular attention to evolutionary biology as a way to explain human nature and behavior. Overall, the thrust of this body of social thought was to embrace humanity's "natural" heritage as animals and to explain human traits as paralleling those of close animal "cousins" who had shared millennia of evolution. Such arguments helped counter claims that American elites—especially men— had become "overcivilized" and were facing nervous extinction. They also marshaled the examples of animals to naturalize gender differences, particularly those favoring male dominance at work and in the home. Boosters of the new brain sciences reworked this discourse, turning attention away from embracing the "natural" and toward tinkering with evolution's gifts through scientifically advanced consumer products such as memory nose sprays, creativity pills, and the like.

Even as psychobiology's visionaries were generating media coverage by promising future miracle drugs, however, the real drug industry was deeply troubled. As we saw in chapters 3 and 4, the best-known wonder drug for the mind, Valium, was famous not as a miracle cure but as an addictive threat to American women. Other drugs in the headlines of the 1970s were similarly notorious, including amphetamines, barbiturates, qualuudes, and a range of painkillers and sleeping pills—all of which seemed to undermine the very idea of wonder drugs and blockbusters.

Nor did the early 1980s begin auspiciously, as researchers found that the brain's own opiates—the basis of the "peptide revolution"—were as addictive as any other kind of narcotic painkiller. No new drugs emerged from their discovery. Psychiatric drugs, like narcotics, continued to be famous as much for their side effects (in particular, their potential to addict) as for their healing power. The new brain sciences were promising, but not much more. Miracle drugs that allowed their users to have only the effects they wanted were still in the future; there was no available technology, nothing people could actually use.

Not surprisingly, drug marketers sought to change the terrain with stories of new wonder drugs that supposedly delivered for the first time on the promises of biological psychiatry. In the late 1970s and early 1980s, for example, advertisers for Tranxene, a would-be successor to Valium, tried to gain an edge by promising that their drug would help women perform "on the job" while creating no "initial euphoria" that might lead to "drug-seeking behavior."[67] But tranquilizers had been too damaged by the Valium panic, at least for the moment; the next blockbuster drug for the mind would have to be found elsewhere.

Antidepressants fared better, ultimately becoming the public face of next-generation miracle drugs and the new brain sciences. As early as 1975, Nathan Kline wrote a piece for *Vogue* featuring a large-print text box declaring that the drugs "do not cause 'craving,' or dependence, nor does tolerance develop as it may with 'uppers' or 'downers.'" The article was titled "No Fun? No Lust? Antidepressant May Bring New Life to Your Life!" and listed new antidepressants by brand name and best qualities.[68] By the 1980s, the appearance of new variants of existing antidepressants (most already in use in Europe) and reports that depression could be measured by relatively simple chemical tests (claims later proved false) made it seem possible that the long heralded miracle drugs were finally at hand.[69] In 1981 and then again in 1986, for example, *New York* magazine profiled "a maverick new generation of antidepressants" that miraculously revived both sex drive and workplace decisiveness in white-collar men and women—giving them exactly what they wanted, and nothing else. By the eve of Prozac's appearance in the late 1980s, journalist Jon Franklin had concluded that "almost no insider" was looking to peptides or schizophrenia for the "first spectacular clinical breakthrough." Instead, "the most promising area of research now is depression."[70] And in the late 1980s, Prozac became famous as the first of these revolutionary products to be actually available—not "just around the corner," but actually at the corner drugstore.

Prozac and the Incorporation of the Brain

Prozac was launched with great marketing fanfare as the first of a new class of antidepressants known as "selective serotonin reuptake inhibitors," or SSRIs. Unlike existing drugs, which raised levels of several monoamines, the SSRIs focused on just one, serotonin. This selectivity had not been a random discovery. Researchers had been looking for just such a drug, guided by a 1970s-era theory that serotonin was the key to depression. While Prozac was not quite as effective as older drugs, it did not appear to produce serious side effects, making it a candidate for treating the milder forms of depression that had been getting so much popular attention in the 1970s and 1980s. It quickly became one of the most widely prescribed medicines in the world, helping catapult antidepressants to prominence as the most important class of psychiatric medicines. Like blockbusters before it, Prozac was prescribed primarily by generalists rather than psychiatrists for a wide range of conditions including depression but also eating disorders, anxiety, sleeplessness, and other classic symptoms from the nervous illness tradition.[71]

Prozac also followed earlier blockbusters in becoming an overnight popular sensation, drawing enough media coverage to become an iconic symbol of its era. The most influential of these popular reports, and a useful touchstone for analysis, was Rhode Island psychiatrist Peter Kramer's 1993 bestseller *Listening to Prozac*. Widely discussed and reviewed in forums from the *New York Review of Books* to *People* magazine, *Listening to Prozac* captured the most dramatic elements of the antidepressant story that had been developing for decades. In it Kramer mixed lessons in the new brain sciences with thoughtful meditations on several patients for whom Prozac not only had eased depression but appeared to have remade their entire personalities. Amazed by his firsthand view of people so fundamentally changed by a chemical intervention, and impressed that there seemed to be no "price" in terms of dopiness or other side effects, Kramer famously concluded that medicine had finally arrived at the point of "cosmetic psychopharmacology" in which identities could be "sculpted" in desirable ways. From that moment on, the up-and-coming blockbuster drug had truly arrived. Anything with "Prozac" on the cover, it seemed, was destined for record sales, and a torrent of books and popular magazine articles deluged the public over the span of a decade, capped by best-selling memoirs such as Elizabeth Wurtzel's *Prozac Nation* (1995) and Lauren Slater's *Prozac Diary* (1998).[72]

Why did Prozac become such a blockbuster? For one thing, as we have seen, the medical ground had been well prepared for it. Partly owing to educational campaigns by drug researchers and pharmaceutical companies, professional medicine had come increasingly to recognize depression, especially mild depression, as a serious scourge, and diagnoses of the illness had been on the rise for decades. The organs of middle-class popular culture too had embraced depression as the new embodiment of the nervous illness tradition, albeit with a more serious aura because of the link to suicide.

Sure enough, Prozac's popularizers worked perfectly within the demographics of the nervous illness tradition. Depression itself might strike people of any social background (as many articles mentioned), but Prozac takers in the news were latter-day neurasthenics in white-collar jobs or careers in the arts. When the *New York Times* profiled the Prozac-spawned "legal drug culture" in 1993, for example, the article began with "six friends, all highly successful professionals in their mid-30s" who met for dinner at a Washington restaurant and discovered that four of them were taking Prozac. *Newsweek* recounted the case of a Chicago woman who took the drug not to cure depression but "to give herself an edge" as a public-relations executive. "No one has tried to count the number of people who fit her profile," the article observed, "but her experience is a parable for the '90s."[73] Aside from Kramer's patients and the occasional celebrity, the most famous Prozac takers were memoirists like Wurtzel, whose book was set during her undergraduate years at Harvard University, and Slater, a well-regarded psychologist and author. (Novelist William Styron used antidepressants other than Prozac for his depression as described in his memoir *Darkness Visible*.)[74]

There was some truth, if perhaps of the self-fulfilling kind, to the white-collar stereotype. Prozac was indeed a drug primarily used by middle-class whites. After all, it was not more effective than earlier drugs; indeed, from the outset even many psychopharmacologists acknowledged that it was no more potent, and possibly less so, than its predecessors. What made Prozac so special was its supposed selectivity and relative lack of serious side effects, which allowed whole new populations of less seriously ill Americans to take it. Not surprisingly, numerous utilization studies revealed that these new populations of pill takers tended to be well-off, white, and female—those with the resources and habits that gave them ready access to prescription medicines, and also those culturally prepared to embrace as "sickness" the kinds of mental states now open to treatment with Prozac.[75]

Depression's new prominence in the nervous illness tradition did not ensure blockbuster success for Prozac, however. Indeed, the Valium panic had left the nervous illnesses hostile territory for wonder drugs of any stripe. But as we have seen, grounds for combating this obstacle had been under preparation as well. For at least a decade before the introduction of Prozac, the cultural spaces of commercialized medicine— popular nonfiction, television, and the health sections of popular magazines—had been circulating stories of revolutionary breakthroughs in brain science and a coming cornucopia of miracle pills. These miracle-pill stories helped Prozac's supporters navigate the cultural politics of the nervous illness tradition and present the antidepressant as a sort of anti-Valium. Feminists and drug critics had condemned tranquilizers as sedating and addictive agents of conformity. Kramer and others described Prozac as a "planfully" constructed tool that allowed precise control of brain function to achieve desirable effects at no psychic cost—the opposite of dope. Mainstream medicine had been criticized for prescribing tranquilizers to keep middle-class housewives content with their lot. Kramer and others described Prozac as a "feminist drug" that made women more assertive and competitive—"supermoms" with careers who laid to rest images of Valium-stoned stay-at-home wives. And finally, Prozac was a brand-name commercially available medicine that women could get through their own volition ("Ask your doctor . . . "). This positioned the drug in the nexus where medicine met commerce in ways that appeared to give pill takers, not doctors, real authority.

It would be hard to overstate the degree to which Prozac became famous as a commercial product—a medicine but also a widely available consumer good that just happened to be available only through physicians. Not coincidentally, the first popular coverage of the drug appeared in *Business Week* in 1988, in an article that praised Prozac's therapeutic promise and speculated that it could "grab 10% or even 20% of the U.S. antidepressant market, estimated at $500 million to $750 million a year."[76] In 1990, *Newsweek* described Prozac as making doctors and patients happy, but the article pointed out that it made drug sellers happy, too: after Prozac "hit the market. . . sales reached $125 million . . . soared to $350 million in 1989 . . . and market analysts expect the drug to pull in $500 million this year and to top $1 billion by 1995," in part because the brand-name drug cost as much as twenty times more than older generics.[77]

For the next decade such commercial facts were omnipresent in public discussions of Prozac and its competitors, and not only in the business and trade journals that honored it as "the product of the year."[78] Combing through popular accounts, it is virtually impossible to find one that did not at least briefly refer to the drug's amazing sales and profitability. Particularly in the stock-obsessed boom years of the 1990s, Prozac's potential as an investment seemed to be everyone's business. Some observers even adopted the language not of medicine but of commerce, identifying patients as "consumers" and the medical delivery system as "the market."[79]

This heightened awareness of Prozac's commercial dimensions was given a boost in the early 1990s by the emergence of direct-to-consumer advertising. The FDA loosened its restrictions on advertising of medications on television and in nonmedical journals, resulting in a flood of promotional materials. Advertising budgets in the drug industry had always been large, and now there were whole new territories to explore. (It is important not to overemphasize this development; as we have seen, pharmaceutical marketing had long been making its way into popular circles. Carefully regulated advertisements were, in all likelihood, more accurate than unregulated public relations campaigns.)[80] While Prozac itself was a latecomer to direct marketing, it did eventually join in. In any case, Prozac, like all drugs, was affected by the implication that physicians were mere middlemen who managed relationships between patients (consumers) and the real providers of treatment (drug manufacturers). As countless advertisements advised, "Ask your doctor about . . ." Peter Kramer almost incidentally raised this issue when one of his patients asked for an increased dose of Prozac because of foreseeable upcoming stresses. He agreed: "Who was I to withhold the bounties of science?"[81]

These accounts of Prozac as a consumer good tended to highlight women's own choices—a big departure from Valium panic stories about women dominated by physicians, drug companies, and ambient sexism. Physicians appeared in the media as sources of technical expertise about Prozac, but with relatively few exceptions they, like Kramer, were merely intermediaries between the narratives' two most important characters: patients (mostly women) and pills. After noting that some physicians were concerned about possible overuse of the easy-to-manage Prozac, for example, the *New York Times* reported that "that has not stopped some patients" and quoted one woman Prozac taker who "has no intention of quitting, ever": "I don't know what the long-term effects of the drug are, but if 20 years from now they say it affects the kidneys or the lungs or whatever, I'm willing to pay the price. Prozac helps me now." A woman interviewed in *New York* magazine described her greatest fear as "that there will be a nuclear war and it'll interrupt my Prozac supply." When *Washington Post* reporter and Prozac taker Tracy Thompson published a story about her personal search for the humans behind the wonder drug, she described the quest as ending not with a physician but a drug researcher, who had no idea what to say to her. Her relationship really was with Prozac itself, not any figure of medical authority. These and countless other reports drove home the basic point that the choice to become a Prozac consumer was ultimately a matter between a patient and the pills, not between a patient and a doctor.[82]

Another way that news coverage of Prozac seemed to be responding to the Valium crisis was in the stories of what happened *after* a woman's choice to take an antidepressant. Prozac, supporters argued, was scientifically crafted to open up new opportunities, not to close them down. Valium and the tranquilizers may have addicted their users, allowing feminists to portray the drugs as trapping and containing women. But Prozac's women consumers ran no such risk, according to boosters. Thanks to science, they were getting the real miracle drug—one that allowed them true choice and control in how they would feel and act.

Building on the monoamine stories that had already been circulating in poplar media, *Listening to Prozac* attributed the medicine's new powers to its precise engineering. Earlier antidepressants, Kramer wrote, were "dirty" drugs that raised brain levels of two or more neurotransmitters (hundreds are known to exist), causing a range of side effects. Prozac, on the other hand, was a "clean" drug that affected only serotonin and had been purposefully selected from thousands of possibilities precisely for this reason. Single-neurotransmitter theories had been batted about for some time, but Prozac was the first and best-known practical application of those theories. The result was not only fewer side effects but, according to Kramer, the wondrous transformation of personality he observed in some of his patients.

This story of technological progress from "dirty" to "clean" medicines was a common theme in popular stories about Prozac. "The scattershot character" of older drugs made them "tricky" to manage, *Newsweek* pointed out in 1990, while Prozac's selectivity made it much easier to use: as one psychiatrist put it, "Instead of using a shotgun you're using a bullet." In 1993 *Time* reported that "Prozac is what scientists call a 'clean' drug. Instead of playing havoc with much of the brain's chemistry, the medication has a very specific effect." Indeed, serotonin itself became somewhat of a minor celebrity as "the mood molecule," repeatedly invoked in magazines from *Mademoiselle* to *Christianity Today* as the chemical basis of happiness. "I didn't know what serotonin was until I found out I didn't have enough of it," began a *Time* magazine essay by novelist Walter Kirn in 1997. "Could a single brain chemical hold the key to happiness, high social status and a nice, flat stomach?" asked *Newsweek* the same year.[83]

Listening to Prozac avoided such direct boosterism, but at the heart of Kramer's book were before-and-after stories of people becoming "better than well." "Tess" won confidence and romantic magnetism; "Sam" found his problematic sexual appetites gone along with his gloom; "Julia" stopped "being a witch" to her family and even found the courage to let her children adopt a dog. However carefully presented, these stories inevitably echoed the miraculous transformations that advertisers promised and that also suffused America's consumer culture generally. In fact, positioning "Tess" and her peers amid more complex analysis paralleled drug advertisements, which featured striking visual images accompanied by denser, text-heavy "brief summaries" with limited indications, side effects, and other warnings. As with the advertisements, the technical weight of the text lent authority and credibility to the central miraculous claims.

Part of Prozac's cultural power was the way it offered boosters a concrete, commercially available example of the coming wonder drugs that brain science was about to deliver. "Doctors . . . are unlocking the mysteries of depression and creating a new science of the mind," *Time* enthused in a 1992 cover story, producing advances "against virtually every affliction to which the human mind is prey." "Prozac is blunt in its effects compared with the anticipated drugs of tomorrow," the *New Republic* asserted. In its review of *Listening to Prozac*, the *Christian Century* agreed that "pharmacology is approaching the point at which psychiatric drugs can reshape personalities." *Newsweek* explained that "the same scientific insights into the brain that led to the development of Prozac are raising the prospect of nothing less than made-to-order, off-the-shelf personalities." Future drugs would target "not so much 'patients' as people who are already functioning on a high level . . . enriching their memory, enhancing intelligence, heightening concentration, and altering for the good people's internal moods." "If you like Prozac," the *New York Times* advised, "you'll love what's coming next"—such as, *Newsweek* suggested in 1997, a "female 'orgasm drug'" and other medicines that might be useful for as much as "a third of the population."[84]

These tales of wondrous new choices offered a different way of thinking about middle-class women's drug use amid Valium-era skepticism. Suddenly, taking a pill might be an empowering act, enabling women to pursue the kinds of life goals that Valium had supposedly taken away. Perhaps the best example of this line of thinking was a strikingly new figure in the public iconography of psychotropic medicines: the "supermom," a middle-class woman who both worked and raised a family. *Time* magazine began a 1993 article with an archetypal example: "Susan Smith has everything going for her. A self-described workaholic, she runs a Cambridge, Massachusetts, real estate consulting company with her husband Charles and still finds time to cuddle and nurture their two young kids . . . What few people know is that Susan, 44, needs a little chemical help to be a supermom: she has been taking the antidepressant Prozac for five years."[85]

This supermom character played a vital role in helping Prozac's boosters present it as the anti-Valium. Prozac, like the tranquilizers, was prescribed for women at rates as much as three times those for men and thus could easily have been criticized as merely the latest "mother's little helper."[86] But the hybrid language of consumer and technological promise cast Prozac in very different terms, replacing narratives of addiction and enforced conformity with liberating tales of choice and empowerment. Prozac's careful engineering, for example, was hailed for producing a drug with no "high" or addiction. As an ad for the drug in *Cosmopolitan* magazine promised, "Prozac is not a happy pill. It's not a tranquilizer: It won't take away your personality. Depression can do that, but Prozac can't."[87]

The supermom character, meanwhile, emphasized that the women who chose to become Prozac consumers did so in pursuit of Betty Friedan's dream of self-fulfillment through activities (especially work) outside the home. Peter Kramer even called Prozac a "feminist" drug because of its power to make women more assertive in their careers and in their personal lives ("No More Ms. Meek," ran a headline about Kramer's book in *U.S. News and World Report*).[88] Prozac's advertisers too tried to play up this aspect of the drug with campaigns such as the one promising "*both* restful nights and productive days." The image of a woman sleeping and then working at an office could not have contrasted more perfectly with the Vivactil advertisement of an earlier decade, which pictured a housewife energized for laundry and then (implied) romance.[89]

Prozac-powered supermoms thus could have it all: white-collar careers, families, good marriages, even a good night's sleep. Prozac's marketing and popularization were perfectly designed to prevent it from becoming the object of feminist activism. It treated a dangerous and potentially fatal illness that was particularly common among women; it was nonaddictive and liberated women from physicians' authority; and it appeared to harmonize with Friedan's demand for women to be more energetic and assertive in pursuing goals outside the home.

Such arguments may have answered the most damning feminist critiques of psychopharmacology, and they probably played a role in warding off the organized and highly visible feminist activism that surrounded tranquilizers and other women's health issues in the 1970s. And there can be no doubt that many women used the drug to pursue their own agendas, as some had also done with tranquilizers. But to call Prozac a

The "supermom" joins housewives in the iconography of psychotropic medicines. *Source:* Prozac ad, *AJP* September 1998, A7. Courtesy of the New York Academy of Medicine Library.

particularly "feminist" drug would be misleading. Indeed, the new Prozac narrative attained such high cultural visibility at least in part because it spoke to a wide range of groups who, for their own reasons, were eager to reject the Valium panic story and the political arguments it had helped circulate. In this way Prozac's celebrity was a response not just to the Valium crisis but to the broader political crises represented by feminism and drug reform. In the more politically conservative climate of the 1980s and 1990s, such a response found many backers.

The Prozac narrative, for example, reemphasized the distinction between medicines and illicit drugs, a line that had become blurred. The most obvious effect of this was to help restore the reputation of the pharmaceutical industry, which had lost prestige and sales in the 1970s but which saw both rebound in the Prozac era. Medicine-drug distinctions were also crucial, however, in another arena: a reinvigorated war against drugs launched during Ronald Reagan's presidency that targeted inner-city users and foreign traffickers with renewed intensity. The new drug war put a decisive end to an era of greater drug toleration, and to the cultural legitimacy of drawing parallels between users of medicines and users of drugs. Among its many other consequences, this had the effect of undermining political interpretations of women's drug use both on the street and from the medicine cabinet.

Antidrug crusaders in the 1980s revived older narratives of drug addicts as violent criminals and marshaled massive new resources behind a militarized campaign to control drugs through punishment rather than treatment.[90] The key figure in the new drug scare was the crackhead, a figure that fully embodied the powerful racial and class politics of American drug policy. Crack was a new smokable form of cocaine, highly potent and relatively cheap, marketed largely to African Americans and Latinos in impoverished inner-city neighborhoods. New drug-war laws imposed penalties for possession of crack that far outweighed those for powder cocaine, which was still used predominantly by whites. The result was a sharp rise in the number of African Americans in prison: they made up less than one-fifth of the nation's drug users but over-two thirds of its incarcerated drug offenders.[91]

Crack was never a widely used drug, and much of the devastation attributed to it could also be explained by the intensification of extreme poverty in America's inner cities owing to broader socioeconomic forces

Depression hurts.

Depression isn't just feeling down. It's a real illness with real causes. Depression can be triggered by stressful life events, like divorce or a death in the family. Or it can appear suddenly, for no apparent reason.

Some people think you can just will yourself out of a depression. That's not true. When you're clinically depressed, one thing that can happen is the level of serotonin (a chemical in your body) may drop. So you may have trouble sleeping. Feel unusually sad or irritable. Find it hard to concentrate. Lose your appetite. Lack energy. Or have trouble feeling pleasure. These are some of the symptoms that can point to depression—especially if they last for more than a couple of weeks and if normal, everyday life feels like too much to handle.

To help bring serotonin levels closer to normal, the medicine doctors now prescribe most often is Prozac.® Prozac isn't a "happy pill." It's not a tranquilizer. It won't take away your personality. Depression can do that, but Prozac can't.

Prozac has been carefully studied for nearly 10 years. Like other antidepressants, it isn't habit-forming. But some people do experience mild side effects, like upset stomach, headaches, difficulty sleeping, drowsiness, anxiety and

Taking Prozac is not doing drugs—the magic is in the serotonin. *Source:* Prozac ad, *Cosmopolitan*, September 1997, 114–15.

Prozac can help.

neryousness. These tend to go away within a few weeks of starting treatment, and usually aren't serious enough to make most people stop taking it. However, if you are concerned about a side effect, or if you develop a rash, tell your doctor right away. And don't forget to tell your doctor about any other medicines you are taking. Some people should not take Prozac, especially people on MAO inhibitors.

As you start feeling better, your doctor can suggest therapy or other means to help you work through your depression. Remember, Prozac is a prescription medicine, and it isn't right for everyone. Only your doctor can decide if Prozac is right for you—or for someone you love. Prozac has been prescribed for more than 17 million Americans. Chances are someone you know is feeling sunny again because of it.

fluoxetine hydrochloride

Welcome back.

 Please see important information on following page.

in the 1980s. But the "crack epidemic" was a successful drug scare none-theless, in part because its central stories helped circulate important po-litical arguments. The rise of political conservatism in the 1980s meant renewed efforts to explain urban poverty as a matter not of missing opportunities but of pathological poor and minority communities. The poor needed moral reform (intact families, church attendance, work ethic, drug abstinence, etc.), not liberal social welfare programs. Crack addiction was a perfect vehicle for dramatizing the pathology of the urban poor, as antidrug crusaders described inner cities as "savage" areas where primitive impulses reigned.[92]

As in the Valium era, drug-war politics were not fully separable from gender politics. And the newly redrawn lines between the street and the medicine cabinet were very significant for at least one population of American women: "crack moms," or pregnant crack users, portrayed by antidrug warriors as the ultimate symbol of depravity. By focusing popu-lar attention on these women and the failure of their "maternal instinct," the drug war figuratively identified these morally damaged individuals as the cause of inner-city devastation. Bad mothers were the real prob-lem, not poverty and lack of opportunities.[93]

Crack moms represented a cultural logic exactly opposite to that of the Valium scare, which had sensationalized addiction to generate sym-pathy and political action rather than horror and punishment. This was not surprising. Valium's critics, after all, had explicitly traded on their whiteness and affluence to gain a hearing for their political arguments about addiction and gender. This left them with little leverage to coun-teract a new cultural campaign to demonize crack moms as the dark mirror image of Prozac's supermoms. While Prozac helped make white and middle-class women more productive on the job and more loving at home, crack helped make inner-city minority women more depraved on the street and more dangerous to others.

Again, such stories may or may not have reflected the realities of using either drug. Prozac, for example, surely made a difference in the lives of many inner-city residents, even though it was prescribed disproportion-ately for well-off white women just as crack was used predominantly by poor nonwhite women.[94] But as a cultural phenomenon, Prozac's "femi-nism" had little to offer women outside the white-collar classes.

Nor were crack moms the only women whose drug use was being

represented as nonpolitical. The white-collar supermoms of the Prozac phenomenon, too, found themselves decoupled from politics in troubling ways. Indeed, Prozac's boosters spoke just as powerfully to the various forces of "postfeminism" as they did to reinvigorated antidrug crusaders. As organized opposition to feminism began to cohere, there were plenty of voices eager to carry the message that the nation's women needed medical treatment, not political change—exactly the opposite of Friedan's original message.

Prozac's supporters tended to emphasize how the drug helped women pursue their career ambitions—to become more competitive, more driven, more confident. Arguing that professional women needed the help of medications, however, raised the possibility that women were naturally unsuited to having careers. The idea that depression was running rampant among career women implied that working might actually be unhealthy for them. Thus when *Fortune* magazine reported in 1995 that almost half the executive women interviewed had taken or had considered taking an antidepressant in the past year, no commentary was necessary to make the point. Women working outside the home needed help, and not of the political variety.[95] It should perhaps be no surprise to discover that in addition to images of productive working women, antidepressant advertisements also continued to feature well-dressed women regaining their passionate devotion to their children and families.[96]

Beyond its implications for women and careers, the Prozac narrative also gave ammunition to a different kind of postfeminism by suggesting that women could get what they needed (or wanted) without political activism. True freedom meant choices—often consumer choices—unconstrained by the demands of politics. But Prozac could not do anything about broader structural impediments to women's freedom that might be causing depression, such as job discrimination, the "double workday," poverty, or (as *Ms.* magazine pointed out) sexual abuse. Resolving women's issues by using a pill deflected feminist arguments by suggesting that women, not society, needed fixing. If, as a popular feminist slogan had claimed, "the personal is political," through Prozac the political became personal. Prozac was "feminist" only insofar as women were the source of their own problems.

None of this is to suggest that Prozac was an "antifeminist" drug, or that it oppressed women. Like Miltown and Valium before it, Prozac surely helped many women pursue their own agendas. As a cultural symbol, however, its popularity was aided by its political usefulness in an era of conservative reaction and retrenchment.

Conclusion: Talking Back to Prozac?

It should come as no surprise to learn that Prozac itself was neither a feminist nor an antifeminist drug. These meanings, like others, emerged from the complex interplay of medical, commercial, and cultural agendas that had been shaping the antidepressant story for decades before the 1980s. By countering the main narratives of the Valium panic, Prozac's supporters helped revive blockbuster medicines, transforming medical practice and reinvigorating the mind-drug industry. They also infused new life into a longstanding arena of cultural contestation: the legal drug cultures of the affluent. Since the nineteenth-century drug cultures of well-to-do matrons using prescription and patent medicines, commercially available mind-altering medications had been important icons in debates about affluent selfhood and gender identity—just as illegal drugs had served as vehicles for circulating disciplinary stories about more marginal populations. In the wonder-drug era, as we have seen, these debates had mostly focused on the danger posed by commercial mind drugs to the "authentic" gender identities of middle-class men and women. Prozac's supporters managed to use this ongoing public conversation for their own purposes, reworking it to form a narrative of liberation through technologically marvelous consumer products.

These popular discussions about Prozac also helped put a consumerist stamp on biological psychiatry and its cousin sociobiology, favoring simple product-oriented models over more complex ways of grounding behavior in biology. The drug itself, as many researchers and theorists have pointed out, offers proof of no such simplistic systems; its effects are as complex and as little understood as those of other medicines for the mind. And yet tales of Prozac and other psychotropic wonder drugs helped frame the new brain science as a relatively simple system that implied a potential for complete mastery of mind and self. Missing from this framing was the ability to imagine a biological basis for selfhood that was not deterministic, that added complexity and mystery rather than taking them away.

As with the Miltown and Valium crazes, however, Prozac did not remain the cultural property of its boosters. Prozac's critics were not, for the most part, feminists; the icon of the supermom was carefully calibrated to defuse feminists' criticisms of wonder drugs as articulated during the Valium panic. But just as addiction researchers helped open tranquilizers to new kinds of public criticism, psychopharmacologists and investigative journalists have begun to uncover disturbing facts about Prozac and its cousin drugs. At the end of the 1980s, for example, Prozac became infamous for allegedly provoking violence or suicide in the people who took it. The Food and Drug Administration soon decided that the claims had been baseless, and a host of lawsuits generally failed to score clear victories. By the time *Listening to Prozac* made the bestseller lists, the issue had been largely forgotten. But researchers such as David Healy and Joseph Glenmullen have recently revisited these claims, leading to new investigations and battles as passionate as any from the Valium addiction saga.[97] Accompanying these investigations has been a torrent of harshly critical literature, reminiscent of the 1970s, exposing pharmaceutical industry corruption and profiteering from ineffective and dangerous drugs—including the new antidepressants, whose wonder-drug luster has tarnished.[98]

It is too early to tell whether this impassioned and increasingly voluminous literature will spark a true drug panic along the lines of the Valium scare. Certainly, new boosterish books about the antidepressants and the endless "new generations" of psychiatric drugs still compete for shelf space with the muckraking literature, and Prozac shows little signs of becoming a negative icon the way Valium did. And yet the ingredients for a panic are all there, as they will always be for blockbuster drugs. As the Prozac phenomenon shows, these drugs are open to such popular crusades because that is exactly the stuff out of which they have been created. They are not built solely by physicians and researchers, but by marketers, public-relations departments, mass-media journalists, and political activists. Their availability as popular symbols helps generate great profits for their manufacturers but also invites others to collaborate in establishing what they mean and how they should be used.

As it had for Miltown and Valium, this contested public process helped turn Prozac into a complex symbol of the promises and perils of seeking a middle-class psychological standard of living—a state of culturally defined self-fulfillment that could be an agent of liberation when achieved by those formerly excluded from it, but also a tool of conformity, tightly bounded by the constraints of medically and commercially defined "health." These cultural dynamics helped give meaning to Prozac and other drugs for achieving well-being, even as they underscored a basic truth: the good life, like all human experiences, is created by human striving—even (or especially) when such striving includes the swallowing of a medicine.

Better Living through Chemistry?

CONTRARY TO WHAT one might initially think, the history of psychiatric drugs is not a story of scientific discoveries and their consequences. To tell this history, one must speak not of molecules—meprobamate, diazepam, fluoxetine hydrochloride—but of the active efforts of many different people who transformed these chemicals into Miltown, Valium, and Prozac. Many Americans—researchers, physicians, and patients; advertisers, lobbyists, and public-relations experts; consumer advocates, antidrug crusaders, feminists, and consumers of popular media—worked to shape the meanings of tranquilizers and antidepressants, and to take advantage of the personal and political opportunities they offered. It was their efforts that turned chemical compounds like meprobamate and fluoxetine hydrochloride into celebrity drugs like Miltown and Prozac, and their efforts too that wrought the social and cultural changes that accompanied the drugs' meteoric rise in the decades since the 1950s.

Physicians and patients have been the most directly involved in these transformations. Beginning with Miltown in the 1950s, new psychiatric drugs offered not only new treatments for age-old problems but also new ways for physicians and patients to interact. The prescription-only pills strengthened physicians' authority, broadening their expertise to cover previously intractable kinds of suffering. They helped psychiatrists lay claim to a scientific footing for their work, promised but never fully delivered by Freudian theories. At the same time, however, the existence of pills whose names and purposes could be discovered by enterprising patients, and whose use ultimately took place outside the doctor's office, also allowed patients to exercise new authority over their own treatment. Thus, for example, a mother and business executive in the 1950s could accept a prescription for Miltown, and then, freed of her debilitating migraines, ignore her doctor's advice to quit her job. As therapeutic tools, therefore, tranquilizers and antidepressants opened new opportunities for both physicians and patients to meet the challenges posed by anxiety, depression, and other forms of very real suffering.

Physicians and patients did not come to embrace the new medicines entirely on their own, however. Indeed, the drugs' very existence stemmed at least in part from pharmaceutical companies' careful gauging of the medical market in search of chances to "do well by doing good." It was Carter Products CEO Henry Hoyt, for example, who saw the commercial possibilities of Frank Berger's weak precursors to meprobamate and invited the researcher to continue developing his new drug while serving as head of Wallace Laboratories in Milltown, New Jersey. And once the medicines had been perfected, it was advertisers who announced their existence to the medical world. Their campaigns gave concrete form to the new medicines' possibilities, expanding medical concepts like "neurosis" and "anxiety" to encompass whole new categories of personal problems. While they made some physicians uncomfortable, commercially defined diagnoses like "environmental depression" or "behavioral drift" clearly helped pave the way for the medicines' widespread use—in part because drug marketers crafted their campaigns to reach popular as well as medical media. Drug advertisers, in other words, helped patients to become consumers actively involved in their own treatment, for better and for worse.

However at odds the two imperatives of curing illness and selling drugs could be, physicians and advertisers shared key assumptions that often made their efforts complementary. Both tended to presume that neurosis, anxiety, and, to a certain extent, depression were all problems that disproportionately struck the white middle classes. For physicians and affluent patients, this was a new chapter in an old tale; at least since the days of neurasthenia, such diagnoses had provided a useful way to define and respond to emotional problems without the stigma of "mental illness." Advertisers, meanwhile, focused on the white-collar world to reinforce their products' cultural connection to science, progress, and health—and to distinguish their medicines from the street drugs associated with nonwhite or marginal populations. Such reasoning also appealed to middle-class cultural critics, who found the reinvigorated medical logic of the nervous illness tradition useful for defining and debating the current needs of America's white-collar classes.

Physicians, advertisers, and patients were not the only ones paying attention to tranquilizers and antidepressants. Addiction researchers, for example, began to report in the 1960s and 1970s that best-selling drugs like Miltown and Valium could produce severe physical dependency in long-term users. Closely tracking the long train of price-fixing and marketing scandals exposed by congressional investigations, consumer advocates denounced tranquilizers as questionably useful or even harmful. Some even likened drug companies to evil pushers who tricked innocents into unnecessary drug dependence. Counterculture youth, meanwhile, defended what they saw as liberating experiments with marijuana and hallucinogens by pointing to the "straight" world's enthusiastic (and, they said, hypocritical) use of prescription sedatives and stimulants. Together with the growing and increasingly voluble ranks of drug treatment specialists, these disparate voices raised a significant challenge to the twentieth century's punitive wars against illicit drugs. Although this challenge never succeeded in overcoming basic distinctions between street and medicine cabinet, it did raise new conflicts within the antidrug coalition. Over the two decades, federal regulators slowly imposed new and often tenuous controls on prescription mood medicines, treading carefully in the face of determined opposition by drug company lobbyists and many physicians.

Perhaps the most important critics of psychiatric drugs have been feminists. Since the 1950s, minor tranquilizers and antidepressants have been used by women at twice the rate of men. As early as 1963, Betty Friedan held up tranquilizers as a powerful symbol of the constraints facing educated housewives—Miltown and Valium, she claimed, were numbing therapies for problems that were, at base, political rather than medical. This argument gained power in the 1970s as the women's health movement coalesced and fears of Valium addiction spread. Ultimately, one wing of the diverse second-wave feminist movement earned a wide audience for their arguments by using the sensational language of antidrug crusaders to portray Valium addiction as an evil consequence of sexist constraints on women's lives. This was a remarkable campaign in some ways, bringing new political morality tales into the drug war and using the spectacle of addiction to push for new opportunities, rather than new punishments, for drug users. But in other ways the Valium panic also hewed closely to drug-war hierarchies. To boost the persuasiveness of their arguments, for example, Valium critics played up the quintessential innocence of middle-class women, often making this point by contrasting them to the degraded figures of street addicts. This helped them earn sympathetic audiences in the mainstream media but significantly narrowed the basis of their critique and decisively cut them off from an already waning broader effort to challenge drug-war practices.

Feminists were not the only ones who knew how to pitch their message so as to catch the interest of mass-market journalists. Drug marketers played up scientific discoveries and revolutionary new treatments while also hinting at near-future possibilities for dispensing with sleep, creating or ending love, controlling memory, and other consumerist fantasies. Addiction research and treatment specialists gained a hearing with sensationalist claims of epidemic drug dependency and horrifying withdrawal symptoms in well-kept middle-class settings. Journalists, long habituated to covering the "drug menace," tried to make sense of the psychotropes by fitting them into established categories of "medicines" or "dope."

Throughout most of this era, the therapeutic use of psychotropes did not depend on any particular theory about why they worked. In the Freudian-dominated 1950s and early 1960s, most physicians—and even most psychopharmacologists—seemed to agree that the pills affected the mechanisms of psychic conflict that Freud had identified as the basis of his "dynamic" psychiatry. For cultural crusaders trying to combat what they called a crisis in masculinity, however, this was a dangerous development. According to them, white-collar men were already "softening" in an era of comfort and conformity; disaster would surely strike if America's best men further quashed their inner dynamism with tranquilizing pills. By stirring up public fascination with Miltown, these masculinist culture warriors were able to gain audiences for their concerns, and to dramatize their proposed solution of returning men to authority at home, at work, and in politics.

The need to defend Miltown from the resulting mini-panic pushed some physicians and advertisers to explore new directions in imagining how psychiatric drugs worked. In the later 1960s and 1970s, this meant borrowing from the emerging science of sociobiology, whose adherents saw humans as animals shaped primarily by their evolutionary heritage. Human social patterns, they argued, could be understood by studying close evolutionary cousins like apes and chimpanzees. Similar reasoning undergirded new claims about psychopharmacology. Tranquilizers, for example, were now described as calming the fight-or-flight reflexes of the primitive hindbrain. This counteracted 1950s-era criticism by casting men's tranquilizer use in simple and hyper-masculine terms, for example depicting men as harboring inner cavemen powered by brute strength. Like sociobiology more generally, the new arguments about tranquilizers added scientific luster to models of powerful, authoritative masculinity.

Such arguments came to fruition in the Prozac era, giving new legitimacy to a kind of reasoning long in the background of the psychopharmacology story. From the beginning, some visionary researchers had promoted the idea that the mind was essentially neurochemical and that all mental and emotional phenomena could be reduced to molecular action. Scientists might not know yet how psychiatric drugs worked, but they were confident that those drugs would ultimately prove to be the key to understanding and controlling the unruly human mind. Armed with technologies to study brain activity, these visionaries played a key role in psychopharmacology in the aftermath of the Valium panic. Advertisers in particular made much use of this story of revolutionary advances in brain science. Marketing campaigns seized on the simplest explanations for how the antidepressants worked, helping make the obscure neurotransmitter serotonin famous as the chemical basis of happiness. The advent of direct-to-consumer advertisements helped these marketing messages spread even further beyond medical circles, providing journalists with an easy narrative for reporting new medical marvels to replace the discredited Valium.

According to its many supporters, Prozac did more than replace Valium as a medicine to ease suffering. It also provided a vehicle for responding to the complex of political challenges that had grown up around tranquilizers. For drug companies, the new technology offered an upbeat, progressive narrative to displace stories of corporate pushers and addicted innocents. The renewed distinctions between medicines and drugs also proved useful to supporters of a greatly intensified war against drugs that deployed militaristic metaphors and policing tactics to target inner-city and foreign drug threats. At the same time, the growing ranks of antifeminists and "postfeminists" in a more conservative era saw much value in the Prozac story, too. Here was a medical and consumerist response to women's unhappiness that seemed to deliver the success that political activism had promised, but without the need for activism. With no addiction or doctorly sexism to blame for it, the widespread use of Prozac could even suggest that feminist goals were unhealthy after all—that careers and autonomy simply made women unhappy. In all cases, the new Prozac narrative allowed physicians, advertisers, and certain groups of cultural campaigners to tell new political morality tales enhanced by the apparent objectivity of science and utopian opportunities of consumerism.

To place human action, rather than drug action, at the heart of this story pulls the history of drugs and medicines out of its isolation and back into contact with other broad developments in postwar America. It enmeshes tranquilizers and antidepressants in the consumer culture, new social movements, drug wars, gender politics, and the other tangled webs of recent history. For efforts to create personal and political opportunities out of new drugs were not unique. They represented only one set of strategies for achieving goals that people also pursued in a variety of other ways, often at the same time.

Perhaps the most basic context for those pursuits was the intensification of America's consumer culture after World War II. The history of psychiatric medicines adds more evidence that consumerism changed in nature as well as scope in the postwar era. Medicines and medical services had been sold in previous eras, and many had even been advertised widely. But in the past, commerce and medicine had been packaged by their boosters as representing different kinds of relationships. Indeed, the professionalization of medicine at the turn of the twentieth century had been, in part, a process of distancing physicians from heavily advertised patent medicines and other aspects of consumer culture. Salesmen were after sales and profits; physicians nobly pursued health and well-being. These boundaries may have been honored more in the breach than in the observance, but they were nonetheless significant; transgressions and exceptions were carefully managed to maintain the proper image.

After World War II, however, professional medicine joined other American institutions in openly embracing advertising, commercialism, and consumerism. Editors in flagship journals like the *Journal of the American Medical Association* may have voiced reservations, but only as an apology for turning over their pages to drug advertisers. There was not supposed to be any shame in it any longer; commercialism was being described not as a threat but as a savior. The mechanics of a consumer economy were built in to the basic functioning of professional medicine. Drug salesmen and advertisers, for example, played (and still play) a crucial role in disseminating information, not just to doctors but also to their patients, educating everyone about a rapidly changing world of therapy. The blockbuster-drug phenomenon, in other words, is not an aberration in an otherwise noncommercial medical system. It is one iconic embodiment of how that system is supposed to work.

Commercialized medicine generated new opportunities for reimagining health and happiness for a wide range of people who manufactured, sold, prescribed, or received new medical treatments. Blockbuster psychiatric drugs distilled these new possibilities with unusual clarity. One of the hallmarks of the modern consumer culture has been claims of miraculous transformations and triumphs. Tranquilizers and antidepressants were nearly archetypal in this regard, promising to satisfy familiar desires for self-confidence, peace of mind, love, happiness, and more. The weight of scientific authority supporting these claims was new, however, as were the size and influence of the social institutions designed to deliver on them. A great deal of cultural firepower had come to be arrayed behind the notion of miracle drugs that could almost literally work magic.

One of the most obvious consequences of the new commercial medicine was to change conceptions of happiness. Advertisers and other message-carriers of the consumer culture had sanctified happiness as a central goal of human striving before, of course. But in the wonder-drug era, professional medicine devoted significant resources to providing happiness, making its pursuit both more accessible and more obligatory, especially for those who could afford access to doctors. As tranquilizers and eventually antidepressants came to be standard appurtenances of middle-class life, happiness itself came to be part of a new psychological standard of living—a disciplinary marker of status that one was obligated to achieve in a variety of ways. Unhappiness in affluent quarters took on some of the qualities of an illness, even as it continued to be interpreted as a sensible attitude on the part of less fortunate populations.

This expectation of happiness could, of course, be liberating for individuals seeking their own path to it. But it could also be limiting, especially when couched in terms of brain chemistry. The notion that pills had the power to provide instant dream fulfillment tended to narrow the meaning of happiness, devaluing the other efforts and human interactions involved in seeking it. Happiness and the drugs used to bring it were increasingly portrayed as issues of molecules, not people.

These miracle-pill stories were carefully designed to bear little resemblance to stories of illegal drugs, but nonetheless brought out underlying similarities between the two. Antidrug crusaders demonized narcotics as uniquely responsible for creating terrifying drug fiends. The individual moral drama of a street addict's struggle with drugs often elbowed out broader social or political narratives that could help explain their troubles. What made this cultural sleight-of-hand possible? The same claims about drugs' magical powers that drove tales of miracle pills: that they could substitute for real experience and produce effects directly.

As Richard DeGrandpre has shown, addiction researchers themselves are considerably less convinced by such reasoning; drug effects on the brain can be reproduced without using drugs, for example, and drugs themselves do not always produce "drug effects."[1] Miracle-drug and demon-drug narratives have been so powerful in part because they subtract the political dynamics from stories about human suffering, happiness, and the strivings that fill the space between them. They highlight individual inner experiences rather than the collective experiences that are the basis of political efforts. So, for example, individual addicts' failure to abstain from drug use became a central narrative in explaining inner-city social collapse in the 1980s, helping to distract attention from deindustrialization, racial segregation, and other contested political narratives.

This kind of logic has been important in the world of legal medicines too, as the accomplishments and limitations of the Valium panic reveal. Feminists successfully reintroduced politics into the story of tranquilizers, forcibly dragging medicines from their isolation as miracle cures into the broader context of postwar gender relations. Women's use of Valium was not separate from other human experiences, they argued; it was just one element of a set of broader social dynamics. To understand the nature of the problem, and to engage it effectively, meant to tackle the broader constraints on women's choices.

Even while challenging the wonder-drug narrative, however, the instigators of the Valium panic accepted its street-drug cousin. They saw Valium as requiring broader political analysis, but saw this as relating to the unique circumstances of middle-class women, not to drug users in general. In fact, media stories about the Valium panic characterized street-drug addicts in familiar terms of individual depravity rather than addressing the broader political and economic constraints they might be facing. This made Valium seem to be an exception—an appalling case where wonder drugs failed—rather than evidence that the moral edifice of the drug war needed to be reexamined. And however tenuous, there were a range of voices making just that broader case in the 1970s, a time when treatment for addicts or even drug legalization gained new political legitimacy, while barbiturates, amphetamines, painkillers, and sleeping pills received new scrutiny and new federal regulations. (Many of the loudest voices, however, did tend to just invert received wisdom, damning all legal substances and praising all illegal ones.) Valium's critics borrowed from but did not fully join this structural critique of the drug war.

One perhaps unexpected consequence of this strategy was that it left few tools for responding to the next wonder-drug phenomenon. In helping stir up fears of tranquilizer addiction, feminists had set their sights on more than Valium. Addiction was a sensationalist vehicle for delivering political arguments about the way social institutions like medicine worked to limit middle-class women's opportunities. But the link between drug use and those broader political arguments, already difficult to make in the midst of America's drug wars, was made even more tenuous by restricting its applicability to Valium and its respectable women users. The politics of Valium use could easily be understood as a problem unique to that particular medicine—the opposite of feminists' goal of showing that Valium was just one small part of a broader system of constraints on women's choices. This meant that the cultural challenges feminists raised could be answered by new wonder-drug boosters, who could (for example) portray Prozac as a next-generation medicine that resolved the Valium dilemma without necessarily engaging political arguments about why so many women were using psychiatric medicines in the first place. But no drug could solve workplace discrimination, "double workdays," sexual assault, or the many other issues confronting American women. And no drug can provide more than the substantial possibility of happiness, a possibility that could become real only as part of life's other personal and political efforts. To imply otherwise was to accept magical thinking about drugs and divorce them from the full contexts in which people take them.

All this is to suggest that distinctions between street and medical drug use need to be challenged aggressively. By this I mean not to deny these distinctions but rather to change their frame of reference to include social and cultural dynamics as well as pharmacology and individual morality. Drug wars, for example, have typically been fought *against* drug users, not *for* them. Indeed, the political and even military resources required for these wars have usually been mobilized by intensifying stereotypes of race or class, or by playing up foreign threats. Such an approach obviously offers little to drug users themselves, and it also creates demonstrably false categories of drug problems (e.g., labeling them as part of the inner city and not the white suburbs). By restoring the full experiences of drug users, and not allowing drug use to erase all other aspects of their lives, we have the potential to wage a very different antidrug campaign—one that uses the tragedy of drug abuse to mobilize the forces of political relief rather than punishment, even if that tragedy unfolds outside the protected quarters of the white middle classes.

Antidrug warriors did not stand alone in building and maintaining divisions between the street and the medicine cabinet. The maintenance of a legal drug culture for the respectable classes served a variety of other purposes, too. Doctors' medicine bags have always been well stocked with pharmacological relief, from opium and morphine through barbiturates and amphetamines to tranquilizers and antidepressants. Linking these "happy pills" to the inner lives of the middle classes has been an important way of protecting them from drug scares. It has also expanded the cultural terrain of the nervous illnesses, which have themselves proved useful constructions. As class-specific maladies with political cures, nervous illnesses allow public discussion of middle-class social problems without fear that the resulting solutions would be taken as applicable to everyone. Thus, for example, Miltown critics called for combating an epidemic of anxiety by reinstating male authority at home and at work, but their medical logic excluded African Americans and other racial minorities who were not yet "advanced" enough to suffer the anxiety in the first place.

One hesitates to apply the notion of "progress" to medicine's growing psychotropic cornucopia, but the development of more drug choices does at the least permit doctors and patients a wider range from which to select the most fitting—and least risky—sort of medicine to treat a particular individual's emotional and mental suffering. As with political sympathy for addicts, it is hard to imagine that this ability to select wisely (or to reject) from many options should be restricted to those able to afford regular visits to the doctor. If we assume that anxiety and depression are at least as common among the poor and marginal as they are among the affluent—if we reject the nervous illness tradition, in other words—and place any value at all in the use of psychiatric drugs to treat them, surely universal access to good medical care would be a reasonable alternative to investing in more drug wars.

But who is to say that making the ever-growing ranks of wonder drugs more widely available would be done beneficially at all? The power of commercialized medicine has only grown enormously since its inception in the postwar era, and little in its makeup suggests the potential for profound commitment to helping those unable to afford insurance or prescriptions. Would it not be likely far worse to have the machinery of commercial medicine aimed at pushing drugs on the nation's most vulnerable populations?

This is no idle question, and it raises issues beyond tranquilizers and antidepressants. Psychiatric medicines are only one of many new commercial technologies like genetics, cloning, cosmetic surgery, and prosthetics that promise to fix human problems directly by changing who we are. On the one hand, these technologies seem wondrous because they offer a shortcut through experience, transforming people regardless of their social, political, or cultural contexts. Like something that defies the law of gravity, they seem to introduce the possibility of entirely new governing rules that allow the impossible. On the other hand, however, these technologies are deeply enmeshed in the very social and economic realities from which they promise escape. They are commercially developed and distributed through mighty institutions that impart their own distinct and often conflicting agendas and meanings. And their promised transformations do not, after all, take place in a vacuum; instead, they create new arenas where economic, political, and cultural struggles take place. These struggles do not bring politics to essentially neutral scientific discoveries; they are the essence of those discoveries. It is such human struggles that we need to plumb to understand how providing or withholding them could create or close down genuine opportunities. And for this task we return to where we began: a story about people as much as about technologies and drugs.

APPENDIX A

Medications Mentioned

Drug Class	*Era of Prominence*	*Brand Name*	*Generic Name*	*Company*
Antianxiety agents (*minor tranquilizers, anxiolytics*)	1950s	Miltown	meprobamate	Carter/Wallace
		Equanil	meprobamate	Am. Home/Wyeth
		Atarax	hydroxyzine	J. B. Roerig
		Vistaril	hydroxyzine pamoate	Pfizer
	1960s/1970s	Valium	diazepam	Roche
		Librium	chlordiazepoxide	Roche
		Serax	oxazepam	Wyeth
		Tranxene	clorazepate dipotassium	Abbott
	1980s/1990	Xanax	alprazolam	Pfizer
Antidepressants (*psychic energizers*)	1950s/early 1960s	Deprol	meprobamate + benactyzine HCl	Carter/Wallace
		Niamid	nialamide	Pfizer
		Marplan	isocarboxazid	Roche
		Nardil	phenelzine dihydrogen	Warner-Chilcott
		Tofranil	imipramine HCl	Geigy
		Elavil	amitriptyline HCl	Merck Sharp & Dohme
	1970s/early 1980s	Aventyl	nortriptyline HCl	Lilly
		Vivactil	protriptyline HCl	Merck Sharp & Dohme
		Pertofrane	desipramine HCl	USV Pharm.
	Late 1980s/ 1990s/2000s	Prozac	fluoxetine HCl	Lilly
		Paxil	paroxetine	Smith Kline & French
		Zoloft	sertraline HCl	Pfizer
		Celexa	citalopram	Forest
		Effexor	venlafaxine	Wyeth

Drug Class	Era of Prominence	Brand Name	Generic Name	Company
Antianxiety/ antidepressant combos	Late 1960s/1970s	Triavil	perphenazine + amitriptyline HCl	Merck Sharp & Dohme
		Sinequan	doxepin HCl	Pfizer
		Etrafon	perphenazine + amitriptyline	Schering
Antipsychotics *(major tran- quilizers, neuroleptics)*	1950s	Thorazine	chlorpromazine	Smith Kline & French
		Serpasil	reserpine	CIBA
		Sandril	reserpine	Lilly
	1960s/1970s	Sparine	promazine HCl	Wyeth
		Trilafon	perphenazine	Schering
		Stelazine	trifluperazine HCl	Smith Kline & French
		Serentil	mesoridazine	Sandoz
		Mellaril	thioridazine	Sandoz
Sedatives/ stimulants	1950s	Butisol	barbital sodium (barbiturate)	McNeil
		Ritalin	methylphenidate HCl (amphetamine-like)	CIBA
		Dexamyl	dextroamphetamine + amobarbital (barb- amp combo)	Smith Kline & French

Prescriptions for Psychiatric Drugs, 1955–2005

☐ Stimulants (amphetamines)

▨ Antidepressants (all nonamphetamine)

▨ Antipsychotics

▨ Sedatives (barbiturates and sleeping pills)

☐ Antianxiety (Miltown, Librium, Valium, other benzodiazepines)

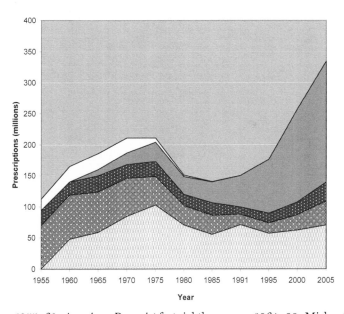

Sources: 1955–68: *American Druggist* fortnightly surveys; 1964–80: Mickey Smith, *A Social History of the Minor Tranquilizers: The Quest for Small Comfort in the Age of Anxiety* (New York: Pharmaceutical Products, 1991); 1964–73: *National Prescription Audit* (Ambler, PA: IMS America); 1974–88: U.S. Department of Commerce, *Drug Utilization in the U.S., Annual Reviews by the Department of Health and Human Services* (Springfield, VA: National Technical Information Service); 1991–2005: Verispan, LLC.

Introduction: Medicine, Commerce, and Culture

1. Paxil ad, *Newsweek*, 17 September 2001, 30–32.

2. Jonathan Metzl, *Prozac on the Couch: Prescribing Gender in the Era of Wonder Drugs* (Durham, NC: Duke University Press, 2003); David Healy, *The Antidepressant Era* (Cambridge: Harvard University Press, 1997), *The Creation of Psychopharmacology* (Cambridge: Harvard University Press, 2002), and *Let Them Eat Prozac: The Unhealthy Relationship between the Pharmaceutical Industry and Depression* (New York: New York University Press, 2006); Edward Shorter, *A History of Psychiatry: From the Era of the Asylum to the Age of Prozac* (New York: John Wiley and Sons, 1997); Mickey Smith, *A Social History of the Minor Tranquilizers: The Quest for Small Comfort in the Age of Anxiety* (New York: Pharmaceutical Products, 1991). Nicolas Rasmussen's book *On Speed: The Many Lives of Amphetamine* (New York: New York University Press, 2008) was published after *Happy Pills* was in press.

3. On the consumer culture see Lizabeth Cohen, *A Consumers' Republic: The Politics of Mass Consumption in Post-war America* (New York: Knopf, 2003); Susan Strasser, Charles McGovern, and Matthias Judt, eds., *Getting and Spending: European and American Consumer Societies in the Twentieth Century* (New York: Cambridge University Press, 1998); Susan Strasser, *ed., Commodifying Everything: Relationships of the Market* (New York: Routledge, 2003); Lawrence Glickman, *A Consumer Society in American History: A Reader* (Ithaca, NY: Cornell University Press, 1999); Robert Weems, *Desegregating the Dollar: African American Consumerism in the Twentieth Century* (New York: New York University Press, 1998).

4. See, e.g., David Musto, *The American Disease: Origins of Narcotic Control*, 3rd ed. (New York: Oxford University Press, 1999); David Courtwright, *Dark Paradise: A History of Opiate Addiction in America* (Cambridge: Harvard University Press, 2001); Nancy D. Campbell, *Using Women: Gender, Drug Policy, and Social Justice* (New York: Routledge, 2000); Joseph Spillane, *Cocaine: From Medical Marvel to Modern Menace in the United States, 1884–1920* (Baltimore: Johns Hopkins University Press, 2000); David Courtwright, *Forces of Habit: Drugs and the Making of the Modern World* (Cambridge: Harvard Uni-

versity Press, 2001); Nayan Shah, *Contagious Divides: Epidemics and Race in San Francisco's Chinatown* (Berkeley: University of California Press, 2001); Caroline Jean Acker, *Creating the American Junkie: Addiction Research in the Classic Era of Narcotic Control* (Baltimore: Johns Hopkins University Press, 2002); Alfred McCoy, *The Politics of Heroin: The Complicity of the CIA in the Global Drug Trade* (New York: Lawrence Hill and Company, 2003); Sarah Tracy and Caroline Jean Acker, eds., *Altering American Consciousness: Essays on the History of Alcohol and Drug Use in the United States, 1800–2000* (Amherst: University of Massachusetts Press, 2004); Curtis Marez, *Drug Wars: The Political Economy of Narcotics* (Minneapolis: University of Minnesota Press, 2004).

5. See, e.g., Edward Shorter, *Bedside Manners: The Troubled History of Doctors and Patients* (New York: Simon and Schuster, 1985); Christopher Callahan and German Berrios, *Reinventing Depression: A History of the Treatment of Depression in Primary Care, 1940–2004* (New York: Oxford University Press, 2005).

6. A number of scholars have begun to pay closer attention to the commercial life of American medicine, particularly in the early twentieth century. For a focus on physicians and commerce, see Andrea Tomes's "Merchants of Health: Medicine and Consumer Culture in the United States, 1900–1940," *Journal of American History*, September 2001, 519–47; "The Great American Medicine Show Revisited," *Bulletin of the History of Medicine* 79 (2005): 627–63; "An Undesired Necessity: The Commodification of Medical Service in the Interwar United States," in *Commodifying Everything*, ed. Strasser, 97–118. Andrea Tone, Elizabeth Siegel Watkins, Nicolas Rasmussen, and David Healy address the commercial development and dissemination of medicines and medical devices. See, e.g., Andrea Tone, *Devices and Desires: A History of Contraceptives in America* (New York: Hill and Wang, 2001); Andrea Tone and Elizabeth Siegel Watkins, eds., *Medicating Modern America: Prescription Drugs in History* (New York: New York University Press, 2007); Elizabeth Watkins, *On the Pill: A Social History of Oral Contraceptives, 1950–1970* (Baltimore: Johns Hopkins University Press, 1998); Nicolas Rasmussen, "The Drug Industry and Clinical Research in Interwar America: Three Types of Physician Collaborator," *Bulletin of the History of Medicine* 79, no. 1 (2005): 50–80; "Steroids in Arms: Science, Government, Industry, and the Hormones of the Adrenal Cortex in the United States, 1930–1950," *Medical History* 46 (2002): 299–324; "Of 'Small Men,' Big Science, and Bigger Business: The Second World War and Biomedical Research in the United States," *Minerva* 40 (2002): 115–46; Healy, *Antidepressant Era, Creation of Psychopharmacology,* and *Let Them Eat Prozac*. Also adding to this scholarship has been the work of business-focused historians such as Louis Galambos, who, e.g., with Jane Eliot Sewell tracked vaccine development at a leading pharmaceu-

tical company in *Networks of Innovation: Vaccine Development at Merck, Sharp and Dohme and Mulford, 1895–1995* (New York: Cambridge University Press, 1997), and Alfred Chandler, *Shaping the Industrial Century: The Remarkable Story of the Evolution of the Modern Chemical and Pharmaceutical Industries* (Cambridge: Harvard University Press, 2005).

7. See Cohen, *Consumers' Republic*.

8. Nicolas Rasmussen argues persuasively that amphetamines were more significant medically (and commercially) as early antidepressants than they have usually been credited. See his "Making the First Anti-depressant: Amphetamines in American Medicine, 1929–1950," *Journal of the History of Medicine and Allied Sciences* 61, no. 3 (2006): 288–323.

9. See, e.g., Dorothy Nelkin, *Body Bazaar: The Market for Human Tissue in the Biotechnology Age* (New York: Crown, 2001).

Chapter One: Blockbuster Drugs in the Age of Anxiety

1. Nathan Hale Jr., *The Rise and Crisis of Psychoanalysis in the United States: Freud and the Americans, 1917–1985* (New York: Oxford University Press, 1995); Gerald Grob, *From Asylum to Community: Mental Health Policy in Modern America* (Princeton: Princeton University Press, 1991); Edward Shorter, *A History of Psychiatry: From the Era of the Asylum to the Age of Prozac* (New York: John Wiley and Sons, 1997).

2. Jonathan Metzl, *Prozac on the Couch: Prescribing Gender in the Era of Wonder Drugs* (Durham, NC: Duke University Press, 2003).

3. Prescription drug makers had always been commercial, of course, even before the war; the change was in scale, scope, and success of marketing and profit making. For prewar commercialism, see especially Nicolas Rasmussen, "The Drug Industry and Clinical Research in Interwar America: Three Types of Physician Collaborator," *Bulletin of the History of Medicine* 79, no. 1 (2005): 50–80; "Steroids in Arms: Science, Government, Industry, and the Hormones of the Adrenal Cortex in the United States, 1930–1950," *Medical History* 46 (2002): 299–324; "Of 'Small Men,' Big Science, and Bigger Business: The Second World War and Biomedical Research in the United States," *Minerva* 40 (2002): 115–46.

4. This and the next two paragraphs on Thorazine's origins are drawn from Judith Swazey, *Chlorpromazine in Psychiatry: A Study of Therapeutic Innovation* (Cambridge: MIT Press, 1974); David Healy, *The Creation of Psychopharmacology* (Cambridge: Harvard University Press, 2002) and *The Psychopharmacologists: Interviews*, vols. 1 and 2 (New York: Chapman and Hall, 1996 and 1998); Frank Ayd and Barry Blackwell, eds., *Discoveries in Biological Psychiatry* (Philadelphia: J. B. Lippincott, 1970); Anne Caldwell, *Origins of Psycho-*

pharmacology from CPZ to LSD (Springfield, IL: Charles C Thomas, 1970); Alfred Burger, "History," in *Psychotherapeutic Drugs, Part I: Principles,* ed. E. Usdin (New York: Marcel Dekker, 1976), 11–59; Walter Sneader, *Drug Discovery: The Evolution of Modern Medicine* (New York: John Wiley and Sons, 1985), esp. 175–91; Sheldon Gelman, *Medicating Schizophrenia: A History* (New Brunswick, NJ: Rutgers University Press, 1999).

5. Joost Meerlo, "Medication into Submission: The Danger of Therapeutic Coercion," *Journal of Nervous and Mental Disorders,* 1955, 353–60, and *The Rape of the Mind: The Psychology of Thought Control, Menticide, and Brainwashing* (New York: World, 1956); Thomas Szasz, "Some Observations on the Use of Tranquilizing Drugs," *AMA Archives of Neurology and Psychiatry,* 1957, 86–92. See also Norman Dain, "Critics and Dissenters: Reflections on 'Antipsychiatry' in the United States," *Journal of the History of the Behavioral Sciences* 25 (1989): 3–25.

6. Interviews giving biographical information about Kline and Ayd appear in Healy, *Psychopharmacologists,* 1:80–110 and 2:223–24, 255–56. For more details on Kline's career, see Nicholas Weiss, "No One Listened to Imipramine," in *Altering American Consciousness: The History of Alcohol and Drug Use in the United States, 1800–2000,* ed. Sarah Tracy and Caroline Jean Acker (Amherst: University of Massachusetts Press), 329–52.

7. David Healy, *The Antidepressant Era* (Cambridge: Harvard University Press, 1997), 65–66, 95–96; F. J. Ayd, "The Early History of Modern Psychopharmacology," *Neuropsychopharmacology* 5 (1991): 71–84.

8. Roland Kuhn, "The Treatment of Depressive States with G22355 (Imipramine Hydrochloride)," *American Journal of Psychiatry,* cited in Healy, *Antidepressant Era,* 43–77.

9. Berger in *Discoveries in Biological Psychiatry,* ed. Ayd and Blackwell, 121–22; Mickey Smith, *A Social History of the Minor Tranquilizers: The Quest for Small Comfort in the Age of Anxiety* (New York: Pharmaceutical Products, 1991); Sneader, *Drug Discovery,* 181–85. See also Andrea Tone, "Tranquilizers on Trial: Psychopharmacology in the Age of Anxiety," in *Medicating Modern America: Prescription Drugs in History,* ed. Andrea Tone and Elizabeth Siegel Watkins (New York: New York University Press, 2007), 158–61.

10. Oscar Schisgall, *Carter-Wallace, Inc.: One Hundred Years, 1880–1980* (Carter-Wallace, 1980), 1–20.

11. For historical development of the pharmaceutical industry, see Alfred Chandler, *Shaping the Industrial Century: The Remarkable Story of the Evolution of the Modern Chemical and Pharmaceutical Industries* (Cambridge: Harvard University Press, 2005); Jeremy Greene, "Attention to 'Details': Etiquette and the Pharmaceutical Salesman in Postwar America," *Social Studies of Science,*

April 2004, 271–92; Harry M. Marks, "Revisiting 'The Origins of Compulsory Drug Prescriptions': Public Health Then and Now," *American Journal of Public Health* 85 (1995): 109–15; John P. Swann, "FDA and the Practice of Pharmacy," *Pharmaceutical History* 36 (1994): 55–70; Jonathan Liebenau, *Medical Science and Medical Industry: The Formation of the American Pharmaceutical Industry* (Baltimore: Johns Hopkins University Press, 1987); Peter Temin, *Taking Your Medicine: Drug Regulation in the United States* (Cambridge: Harvard University Press, 1980), 18–64; Walter Measday, "The Pharmaceutical Industry," in *The Structure of American Industry,* ed. Walter Adams, 4th ed. (New York: Macmillan, 1971), 156–57. But see also Rasmussen, "Making the First Antidepressant" and "The Drug Industry and Clinical Research in Interwar America," which show that even ethical firms were involved in marketing and competition before World War II.

12. Measday, "Pharmaceutical Industry," 157–61; Temin, *Taking Your Medicine,* 64–82; Chandler, *Shaping the Industrial Century.*

13. Figures from Federal Trade Commission, Securities Exchange Commission, and *Moody's Industrial Manual* collected by Paul Rand Dixon, counsel and staff director for Tennessee senator Estes Kefauver, and printed in U.S. Senate, Committee on the Judiciary, Subcommittee on Antitrust and Monopoly, *Administered Prices in the Drug Industry* (Washington, DC: GPO, 1959–61), 8916. Tranquilizer market figures from U.S. Senate, Committee on Government Operations, Subcommittee on Reorganization and International Organizations, *Interagency Coordination in Drug Research and Regulation* (Washington, DC: GPO, 1964), 1409–10, also reported in *F-D-C Reports "Pink Sheet,"* April 30, 1956, 17. See also Chandler, *Shaping the Industrial Century.*

14. Measday, "Pharmaceutical Industry," 162, 175–78; Temin, *Taking Your Medicine,* 82–86.

15. Vance Packard, *The Hidden Persuaders* (New York: D. McKay, 1957); Lizabeth Cohen, *A Consumers' Republic* (New York: Knopf, 2003); Stephen Fox, *The Mirror Makers: A History of American Advertising and Its Creators* (New York: Morrow, 1984).

16. Richard McFadyen, "The FDA's Regulation and Control of Antibiotics in the 1950s: The Henry Welch Scandal, Félix Martí-Ibáñez, and Charles Pfizer and Co.," *Bulletin of the History of Medicine,* Summer 1979, 159–69; U.S. House, Committee on Government Operations, Subcommittee on Legal and Marketing Affairs (1958), *False and Misleading Advertising (Prescription Tranquilizing Drugs),* 85th Cong., 2nd sess. (Washington, DC: GPO, 1958). The "Kefauver hearings"—U.S. Senate, Committee on the Judiciary, Subcommittee on Antitrust and Monopoly (1959–61), *Administered Prices in the Drug Industry (Tranquilizers)*—did ultimately result in an important new law, but not the one he was

aiming for: because of the Thalidomide tragedy in Europe, where hundreds of babies were born with serious birth defects after their mothers took a supposedly safe tranquilizer, Congress passed a bill mandating more rigorous testing of safety and effectiveness for new drugs—but did nothing about drug costs, and little about marketing practices. Other federal hearings included U.S. Senate, Select Committee on Small Business, Subcommittee on Monopoly (1969), *Competitive Problems in the Drug Industry,* 91st Cong., 1st sess. (Washington, DC: GPO, 1969); U.S. Senate, Select Committee on Small Business, Subcommittee on Monopoly (1971), *Advertising of Proprietary Medicines, Part 2: Mood Drugs (Sedatives, Tranquilizers, and Stimulants,* 92nd Cong., 1st sess. (Washington, DC GPO, 1971); U.S. Senate, Committee on Labor and Public Welfare, Subcommittee on Health (1974), *Examination of the Pharmaceutical Industry,* 93rd Cong., 2nd sess. (Washington, DC: GPO, 1974).

17. Schisgall, *Carter-Wallace, Inc.,* 26–37.

18. Tone, "Tranquilizers on Trial," 159–60.

19. *The Wyeth Story: The History of a Major American Pharmaceutical Firm* (Wyeth Laboratories, n.d.); *Looking Back: A Pictorial Essay on the History of Wyeth-Ayerst* (Wyeth Laboratories, n.d.).

20. Estimates by American Home Products, Inc., printed in U.S. Senate, *Administered Prices in the Drug Industry,* 10076.

21. Figures on Carter sales are from William Latourette, "Less Tranquility: Competition in Mental Health Drugs Grows More Intense," *Barron's,* March 31, 1938, 13; "Growth Trend For Two Drug Firms," *Financial World,* April 2, 1958, 13; Ira Cobleigh, "The Tranquil Rise of Carter Products, Inc.," *Commercial and Financial Chronicle,* May 8, 1958, 4; Louis Rolland, "Carter's Tranquility," *Financial World,* July 2, 1958, 22–23; "Tranquilizers, New Lines Stimulate Strong Growth by Carter Products," *Barron's,* December 13, 1958, 21–23; Robert Newton, "Calling Attention . . . to Changing Patterns in the Drug Industry," *Magazine of Wall Street,* October 8, 1960, 88–90.

22. Sparine, a major tranquilizer, accounted for another 5 percent of sales. Profit figures for American Home from U.S. Senate, *Administered Prices in the Drug Industry,* 8910, 8951; Latourette, "Less Tranquility," 13; Newton, "Calling Attention," 88–90.

23. U.S. Senate, *Examination of the Pharmaceutical Industry,* 747–55. For information about detail men, see Greene, "Attention to 'Details.'" Also the testimony of former Roche salesman Charles Brannan and the rebuttal by Roche Vice President of Marketing and Promotion Irwin Lerner in U.S. House, Select Committee on Narcotics Abuse and Control, *Abuse of Dangerous Licit and Illicit Drugs—Psychotropics, Phencyclidine (PCP), and Talwin,* 95th Cong., 2nd sess.,

August 10, 1978 (Washington, DC: GPO, 1979), 121–49, 168–95; internal Roche sales documents reprinted in U.S. Senate, Committee on Labor and Human Resources, Subcommittee on Health and Scientific Research, *Examination on the Use and Misuse of Valium, Librium, and Other Minor Tranquilizers*, 96th Cong., 1st sess., September 10, 1979 (Washington, DC: GPO, 1979), 173–79.

24. According to a report by Peat, Marwick, Mitchell, and Co. for Senator Kefauver's investigation, *JAMA*'s advertisement revenue nearly doubled from $3.8 million in 1955 to $6.75 million in 1958; "Senate Anti-trust and Monopoly, Drugs," accession 71A 5170, record group 46, box 19, National Archives Building, Washington, DC. AMA advertising policy changes followed the advice of marketing consultants whose report, *A Study of Medical Advertising and the American Physician: An Opinion Survey Made for the American Medical Association* (Chicago: Ben Gaffin and Associates, August 31, 1953), can be found in "Senate Anti-trust and Monopoly, Drugs," accession 71A 5170, record group 46, box 19.

25. Atarax ads, *JAMA*, May 5, 1956, 89, and May 26, 1956, 30–31; Meprotab ad, *JAMA*, November 16, 1957, 97.

26. Equanil ad, *JAMA*, December 21, 1957, 29.

27. Carter's conference printed in *Journal of Neuropsychiatry*, September–October 1964, 386–489. One participant recalled his experience there in an editorial in the *Journal of the Medical Society of New Jersey*, July 1964: "If any of the speakers pointed out the disadvantages of living on tranquilizers, his remarks escaped the notice of your observer" (246). Roche's conference addressed in U.S. Senate, *Examination on the Use and Misuse of Valium, Librium, and Other Minor Tranquilizers*, 210–89. Questioned in Congress about Valium's prominent role in the conference, a Roche spokesman protested innocence: "You're a terrible cynic."

28. Fox, *Mirror Makers*.

29. Thomas Whiteside, "Getting There First with Tranquility," *New Yorker*, May 3, 1958, 117.

30. William Castagnoli, *Medicine Avenue: The Story of Medical Advertising in America* (Huntington, NY: Medical Advertising Hall of Fame, 1999), 15–16.

31. Milton Moskowitz, "Librium: A Marketing Case History," *Drug and Cosmetic Industry* 87 (1960): 460–61, 566–67, and "Librium Becomes No. 1 Tranquilizer, Though Outspent by Rival Advertiser," *Advertising Age*, September 12, 1960, 3, 66–68, 72; "New Way to Calm a Cat," *Life*, April 18, 1960, 93–95.

32. See *American Druggist*'s "Prescription Trends" for 1955–65. American Home Products' internal audit showed meprobamate with 70% of all tranquilizer prescriptions in 1956; U.S. Senate, *Administered Prices in the Drug Industry*, 10076.

33. *American Druggist,* fortnightly prescription surveys, 1955–65.

34. "Barbiturates Are Sociable; That Explains Why Rxs Increase Despite Tranquilizers," *American Druggist,* June 2, 1956, 32+; "Growth: In This Year, U.S. Doctors Will Prescribe 49 Million Rxs Containing Tranquilizers," *American Druggist,* June 18, 1956, 33+.

35. Hans Selye, *The Stress of Life* (New York: McGraw-Hill, 1956); Raymond Cattell and Ivan Scheier, *The Meaning and Measurement of Anxiety* (New York: Ronald, 1961); Raymond Cattell, in "Symposium on Anxiety and a Decade of Tranquilizer Therapy: Psychological Definition and Measurement of Anxiety," *Journal of Neuropsychiatry,* September–October 1964, 396–402. For the history of professional psychology, see Ellen Herman, *The Romance of American Psychology: Political Culture in the Age of Experts* (Berkeley: University of California Press, 1995), and Donald Napoli, *Architects of Adjustment: The History of the Psychological Profession in the United States* (Port Washington, NY: Kennikat, 1981).

36. Lowell Selling, "Clinical Study of a New Tranquilizing Drug," *JAMA,* April 30, 1955, 1594–96; Joseph Borrus, "Study of Effect of Miltown (2-methyl-2-n-propyl-1,3-propanediol dicarbamate) on Psychiatric States," *JAMA,* April 30, 1955, 1596–98.

37. *Miltown: The Tranquilizer with Muscle Relaxant Action* (New Brunswick, NJ: Wallace Laboratories, 1958). One likely reason for the profusion of uses, as Nicolas Rasmussen has shown with amphetamines and other medicines, was that drug companies actively solicited "physician collaborators" who could find and provide evidence for new uses for their products. See Rasmussen, "The Drug Industry and Clinical Research in Interwar America" and "Making the First Antidepressant."

38. John Krantz and C. Jelleff Carr, *The Pharmacologic Principles of Medical Practice* (Baltimore: Williams and Wilkins, 1958), 707. See also Helen Kaplan, "Tranquilizers in the Office Practice of Medicine," *New York State Medical Journal,* August 1, 1959, 2871: "Since the advent of antibiotics there probably has been no advance in medicine that has had such a uniquely wide and revolutionary application as that of the tranquilizing drugs."

39. Grob, *From Asylum to Community,* esp. 93–102.

40. Edward Shorter, *Bedside Manners: The Troubled History of Doctors and Patients* (New York: Simon and Schuster, 1985), 140–78; Grob, *From Asylum to Community;* Eva Moskowitz, *In Therapy We Trust: America's Obsession with Self-Fulfillment* (Baltimore: Johns Hopkins University Press, 2001).

41. Committee on Nomenclature and Statistics of the American Psychiatric Association, *Diagnostic and Statistical Manual: Mental Disorders* (Washington, DC: American Psychiatric Association, Mental Hospital Service, 1952), 31–34.

42. Noah Dixon, "Meprobamate, a Clinical Evaluation," *Annals of the New York Academy of Sciences*, May 9, 1957, 775. For the history of psychosomatic medicine, see Patricia Jasen, "Malignant Histories: Psychosomatic Medicine and the Female Cancer Patient in the Postwar Era," *Canadian Bulletin of Medical History* 20 (2003): 265–97; Edward Shorter, *From the Mind into the Body: The Cultural Origins of Psychosomatic Symptoms* (New York: Free Press, 1994).

43. See Leo Srole at al., *Mental Health in the Metropolis* (New York: McGraw-Hill, 1962); August Hollingshead and Frederick Redlich, *Social Class and Mental Illness: A Community Study* (New York: Wiley, 1958); Benjamin Pasamanick, ed., *Epidemiology of Mental Disorder: A Symposium Organized by the American Psychiatric Association* (Washington, DC: American Association for the Advancement of Science, 1959); Arnold Rose et al., *Mental Health and Mental Disorder: A Sociological Approach* (New York: Norton, 1955); Paul Hoch and Joseph Zubin, eds., *Comparative Epidemiology of the Mental Disorders* (New York: Grune and Stratton, 1961); Derek Phillips, "The 'True Prevalence' of Mental Illness in a New England State," *Community Mental Health Journal*, Spring 1966, 35–40. See also Gerald Grob, "The Origins of American Psychiatric Epidemiology," *American Journal of Public Health*, March 1985, 229–36.

44. Lea Associates, Inc., *The National Disease and Therapeutic Index, 1962–1965* (Ambler, PA: Lea Associates, 1962–76).

45. M. Ralph Kaufman and Stanley Bernstein, "A Psychiatric Evaluation of the Problem Patient: Study of a Thousand Cases from a Consultation Service," *JAMA*, January 12, 1957, 108–11.

46. See, e.g., Michael Shepherd, *Psychiatric Illness in General Practice* (London: Oxford University Press, 1966); Richard Finn and Paul Huston, "Emotional and Mental Symptoms in Private Medical Practice: A Survey of Prevalence, Treatment, and Referral in Iowa," *Journal of the Iowa Medical Society* 56 (1966): 673–77; Peter Hesbacher et al., "Psychotropic Drug Prescription in Family Practice," *Comprehensive Psychiatry*, September–October 1976, 607–15; Shorter, *From the Mind into the Body.*

47. Cited in George Albee, *Mental Health Manpower Trends* (New York: Basic Books, 1959), 301.

48. *Mental Health Manpower Trends* (New York: Basic Books, 1959), 301–19. For an excellent discussion of the Commission and its impact on American psychiatry, see Grob, *From Asylum to Community*, 181–208.

49. Nathan Kline, "Psychopharmaceuticals: Uses and Abuses," *Postgraduate Medicine*, May 1960, 621. See also, e.g., Frank Orland, "Use and Overuse of Tranquilizers," *JAMA*, October 10, 1959, 633; L. J. Meduna, "Use of Drugs in the Treatment of Neuroses and in the Office Management of Psychoses," *Modern Medicine*, November 15, 1957, 212; Ronald Koegler, "Drugs, Neurosis, and the

Family Physician," *California Medicine*, January 1965, 5–81; Jackson Smith, Merritt Foster, and Lester Rudy, "Tranquilizers and Energizers—or Neither?" *Postgraduate Medicine*, April 1963, 350.

50. Miltown ad, *JAMA*, November 5, 1955, 31; Miltown ad, *JAMA*, November 3, 1956, 27. Not to be outdone, Equanil's advertisers recommended their drug for "the common problems of everyday practice"; Equanil ad, *JAMA*, April 26, 1958, 108.

51. Miltown ad, *American Journal of Psychiatry*, January 1958, front cover.

52. Atarax ad, *JAMA*, May 4, 1957, 79.

53. Equanil ads, *JAMA*, November 17, 1956, 101; November 24, 1956, 45; September 15, 1956, 89; September 22, 1956, 53; October 13, 1956, 61; October 20, 1956, 79.

54. Equanil ad, *JAMA*, February 4, 1956, 76–77.

55. Kaplan, "Tranquilizers in the Office Practice of Medicine," 2874–75.

56. Chauncey Leake, "Introduction to Symposium on Anxiety and a Decade of Tranquilizer Therapy," *Journal of Neuropsychiatry*, September–October 1964, 387. Robert Schmitt, writing in the analytic *Psychiatric Quarterly*, suggested that "the tranquilizing agents have been effective because of their influence upon the basic force in the production of mental symptomatology—anxiety"; "The Psychodynamics of the Tranquilizing Agents," *Psychiatric Quarterly* 31 (1957): 29. See also G. J. Sarwer-Foner, "On the Mechanisms of Action of Neuroleptic Drugs: A Theoretical Psychodynamic Explanation," *Recent Advances in Biological Psychiatry*, June 7–9, 1963, 217–32.

57. Herbert Berger, "Management of the Neuroses by the Internist and General Practitioner," *New York State Medical Journal*, June 1, 1956, 1783.

58. See, e.g., Leo Hollister, "Tranquilizing Drugs and the Generalist," *Medical Times*, October 1956, 1021–66; Joseph Fazekas et. al., "Ataractics in General Practice," *GP*, December 1956, 75–81; Walter Tucker, "The Place of Miltown in General Practice," *Southern Medical Journal*, September 1957, 1111–14; L. J. Meduna, "Use of Drugs in the Treatment of Neuroses and in the Office Management of Psychoses," *Modern Medicine*, November 15, 1957, 212; Arthur Marshall, "Psychiatric Problems in General Practice: Psychopharmacology of Drugs," *California Medicine*, May 1958, 345–47; Julius Michaelson, "The General Practitioner's Role in the Control of Anxious and Neurotic Patients," *Journal of Neuropsychiatry*, September–October 1964, 440–41; Koegler, "Drugs, Neurosis, and the Family Physician."

59. Ben Eisenberg, "Role of Tranquilizing Drugs in Allergy," *JAMA*, March 16, 1957, 936–37.

60. One psychiatry textbook, for example, described a 45-year-old stockbroker suffering from insomnia and claustrophobia who was cured after Miltown

enabled him to "realize and work through the underlying dynamic conflicts" in previously fruitless twice-weekly sessions. These "dynamic conflicts" turned out to be the classically psychoanalytic problem of competitive feelings toward his son, who had just begun puberty. Frederic Flach and Peter Regan, *Chemotherapy in Emotional Disorders: The Psychotherapeutic Use of Somatic Treatments* (New York: McGraw-Hill, 1960), 132–39. For a useful interpretation of such Freudian continuities, see Metzl, *Prozac on the Couch*.

61. Lawrence Kolb, "Anxiety and the Anxiety States," *Journal of the Chronic Diseases*, March 1959, 210; Kolb and Arthur Noyes, *Modern Clinical Psychiatry* (Philadelphia: W. B. Saunders, 1958, 1963); Arthur Marshall, "Psychiatric Problems in General Practice: Psychopharmacology of Drugs," *California Medicine*, May 1958, 345.

62. See, e.g., Mortimer Ostow, *Drugs in Psychoanalysis and Psychotherapy* (New York: Basic Books, 1962), 1–11; Karl Rickels, "The Use of Psychotherapy with Drugs in the Treatment of Anxiety," *Psychosomatics*, March–April 1964, 111–12.

63. Brian Campden-Main and Thelma Campden-Main, "American Psychiatry: Current Practices and Views in Psychopharmacology," *Diseases of the Nervous System*, March 1962, 135–40; Max Hayman and Keith Ditman, "Influence of Age and Orientation of Psychiatrists on Their Use of Drugs," *Comprehensive Psychiatry*, June 1966 [survey completed in 1962], 152–65.

64. Harold Himwich, "Psychopharmacologic Drugs," *Science*, January 1958, 61.

65. Joseph Borrus, "Meprobamate in Psychiatric Disorders," *Medical Clinics of North America*, March 1957, 327–37; Louis Linn in Ralph Kaufman et al., "Clinical Conference: Early Recognition and Management of Psychiatric Disorders in General Practice," *Journal of the Mount Sinai Hospital*, March–April 1958, 137–59.

66. Robert Ferber and Hugh Wales, "The Effectiveness of Pharmaceutical Advertising: A Case Study," *Journal of Marketing*, April 1958, 398–407; Raymond Bauer and Lawrence Wortzel, "Doctor's Choice: The Physician and His Sources of Information about Drugs," *Journal of Marketing Research*, February 1966, 40–47; Lawrence Linn and Miltown Davis, "Physicians' Orientation toward the Legitimacy of Drug Use and Their Preferred Source of New Drug Information," *Social Science and Medicine*, 1972, 199–203; Russell Miller, "Prescribing Habits of Physicians: A Review of Studies on Prescribing of Drugs," *Drug Intelligence and Clinical Pharmacy*, November 1973, 493–500, December 1973, 557–64, February 1974, 81–91; Colman Herman and Christopher Rodowskas, "Communicating Drug Information to Physicians," *Journal of Medical Education*, March 1976, 189–95; Mickey Smith, "Drug Product Advertising and

Prescribing a Review of the Evidence," *American Journal of Hospital Pharmacy* 34 (November 1977): 1208–24.

67. Sam Shapiro and Seymour Baron, "Prescriptions for Psychotropic Drugs in a Noninstitutional Population," *Public Health Reports*, June 1961, 481–88, and "Use of Psychotropic Drug Prescriptions in a Prepaid Group Practice Plan," *Public Health Reports*, October 1962, 871–78. These proportions remained remarkably constant over time, with a 1970 national survey indicating, again, that slightly over one-third of minor tranquilizers were prescribed for psychiatric disorders. New was the increased importance of prescribing for "senility," which accounted for a full one-sixth of all minor tranquilizer use. Hugh Parry et al., "National Patterns of Psychotherapeutic Drug Use," *AMA Archives of General Psychiatry*, June 1973, 769–83.

68. Herman Dickel and Henry Dixon, "Inherent Dangers in Use of Tranquilizing Drugs in Anxiety States," *JAMA*, February 8, 1957, 422–26; AMA Council on Drugs, "Potential Hazards of Meprobamate," *JAMA*, July 20, 1957, 1332–33; Leon Powell et. al., "Acute Meprobamate Poisoning," *New England Journal of Medicine*, October 9, 1958, 716–18; Charles McKown et al., "Overdosage Effects and Danger from Tranquilizing Drugs," *JAMA*, August 10, 1963, 425–30.

69. Harry J. Marks, *The Progress of Experiment: Science and Therapeutic Reform in the United States, 1900–1990* (New York: Cambridge University Press, 1997); Healy, *Antidepressant Era*, chap. 2, "Other Things Being Equal." In many studies, physicians had continued to provide talk therapy as well as Miltown, and so, as one critic pointed out, "the glowing results of the clinical trial may be, at least in part, a glowing tribute to [their] ability as psychotherapist[s.]" Victor Laties and Bernard Weiss, "A Critical Review of the Efficacy of Meprobamate (Miltown, Equanil) in the Treatment of Anxiety," *Journal of Chronic Diseases*, June 1958, 502.

70. "Meprobamate," *Medical Letter on Drugs and Therapeutics*, January 23, 1959, 3–4. See also T. F. Rose, "The Use and Abuse of the Tranquilizers," *Canadian Medical Association Journal*, January 15, 1958, 144–48; Dale Friend, "Current Concepts in Therapy: Tranquilizers, III: Meprobamate, Phenaglycodol and Chlordiazepoxide," *New England Journal of Medicine*, April 27, 1961, 872; Harry Beckman, *Pharmacology: The Nature, Action, and Use of Drugs*, 2nd ed. (Philadelphia: W. B. Saunders, 1961).

71. David Greenblatt, "Meprobamate: A Study of Irrational Drug Use," *American Journal of Psychiatry*, April 1971, 1297–1303. Miltown had fallen so thoroughly that in 1964 a physician's guide to psychiatric drugs distributed by a new psychiatric wing of a New York hospital described it as both more toxic and more

addictive than the barbiturates—a claim for which there appears to have been no evidence; Arthur Shapiro, "Rational Use of Psychopharmaceutic Agents," *New York State Journal of Medicine*, May 1, 1964, 1090.

72. Louis Goodman and Alfred Gilman, *The Pharmacological Basis of Therapeutics: A Textbook of Pharmacology, Toxicology, and Therapeutics for Physicians and Medical Students*, 2nd ed. (New York: Macmillan, 1955) and 3rd ed. (New York: Macmillan, 1965).

73. See the special section on "benzodiazepines" in U.S. Department of Commerce, *Drug Utilization in the U.S.—1985: Seventh Annual Review* (Springfield, VA: National Technical Information Service, 1986). Smith, *Social History of the Minor Tranquilizers*, 31–32, also includes a chart of prescription patterns for all the various psychiatric drugs up to 1980.

74. Leo Sternbach, "The Discovery of Librium," *Agents and Actions* 2, no. 4 (1972): 193–96; Smith; *Social History of the Minor Tranquilizers;* B. I. K., "The Father of Mother's Little Helpers," *U.S. News and World Report*, December 27, 1999, 58.

75. Winston Burdine, "Diazepam in General Psychiatric Practice," *American Journal of Psychiatry*, December 1964, 589–92.

76. Joseph Tobin and Nolan Lewis, "New Psychotherapeutic Agent, Chlordiazepoxide," *JAMA*, November 5, 1960, 1242–49.

77. Ibid.; Burdine, "Diazepam in General Psychiatric Practice"; Moke Williams, "Clinical Impressions on the Use and Value of Chlordiazepoxide in Psychiatric Practice," *Southern Medical Journal*, August 1961, 822–26; S. Foster Moore, "Therapy of Psychosomatic Symptoms in Gynecology: An Evaluation of Chlordiazepoxide," *Current Therapeutic Research*, May 1962, 249–57; M. Vilkin and J. Lomas, "Clinical Experience with Diazepam in General Psychiatric Practice," *Journal of Neuropsychiatry*, August (suppl.) 1962, 139–44; George Constant and Frank Gruver, "Preliminary Evaluation of Diazepam in Psychiatric Disorders," *Psychosomatics*, March–April 1963, 80–84; Julian Love, "Diazepam in the Treatment of Emotional Disorders," *Diseases of the Nervous System*, November 1963, 674–77; Robert Rathbone, "The Role of a Psychotherapeutic Drug in Internal Medicine," *Medical Times*, December 1963, 1186–91; Henry Cromwell, "Controlled Evaluation of Psychotherapeutic Drug in Internal Medicine," *Clinical Medicine*, December 1963, 2239–44; G. H. Aivazian, "Clinical Evaluation of Diazepam," *Diseases of the Nervous System*, August 1964, 491–96; Stanley Dean, "Diazepam as an Adjuvant in Psychotherapy," *American Journal of Psychiatry*, October 1964, 389–90; Wilfred Dorfman, "Recent Advances in Psychopharmacology," *Diseases of the Nervous System*, November 1963, 694–97; Robert Burnett and Robert Holman, "Experience with Valium in General Prac-

tice," *Medical Times*, January 1965, 56–60; George Sprogis, "Control of Anxiety/ Depression Reactions in General Practice," *Current Therapeutic Research*, October 1966, 490–93.

78. Lothar Kalinowsky and Paul Hoch, *Somatic Treatments in Psychiatry*, 3rd ed. (New York: Grune and Stratton, 1961), 96.

79. Angus Bowes, "The Role of Diazepam (Valium) in Emotional Illness," *Psychosomatics*, September–October 1965, 336–40.

80. Librium ad, *American Journal of Psychiatry*, April and May 1960, January 1962.

81. "Librium and Valium," *Medical Letter on Drugs and Therapeutics*, October 3, 1969, 81–84; "Diazepam as a Muscle Relaxant," *Medical Letter on Drugs and Therapeutics*, July 6, 1973, 57–58; "Antianxiety Drugs in Organic and Functional Syndromes," *Medical Letter on Drugs and Therapeutics*, December 8, 1972, 93–94.

82. See also, e.g., the "educational" material provided by the pharmaceutical industry public-relations organization Health News Institute: "Tranquilizer Drugs—An Identification" (New York: Health News Institute, January 15, 1960), in "Senate Anti-Trust and Monopoly, Drugs," accession 71a 5170, record group 46, box 6, National Archives Building, Washington, DC; or the on-radio interviews given by William Apple, the executive secretary of the American Pharmaceutical Association, and Francis Brown, president of Schering Corporation, on "American Forum of the Air" (Westinghouse Broadcasting Company, February 1, 1960), in "Senate Anti-Trust and Monopoly, Drugs," accession 71a 5170, record group 46, box 19.

83. Whiteside, "Getting There First with Tranquility"; "Don't-Give-a-Damn Pills," *Time*, February 27, 1956, 98 (includes picture of "Miltown" Berle); "Pills vs. Worry—How Goes the Frantic Quest for Calm in Frantic Lives?" *Newsweek*, May 21, 1956, 68–70; "Happiness by Prescription," *Time*, March 11, 1957, 59.

84. "Tranquil Pills Stir up Doctors," *Business Week*, June 28, 1958, 28–30.

85. *Time*, March 15, 1963, cited in U.S. Senate, *Interagency Coordination in Drug Research and Regulation*, 1275–79.

86. Francis Dickel, "The Tranquilizer Question," *Fortune*, May 1957, 164; Arthur Gordon, "Happiness Doesn't Come in Pills," *Reader's Digest*, January 1957, 60–61 (condensed from *Woman's Day*, January 1957). See also Alek Rozental, "The Strange Ethics of the Pharmaceutical Industry," *Harper's*, May 1960, 78–79.

87. "New Way to Calm a Cat," *Life*, April 18, 1960, 93–95; "Tranquil but Alert," *Time*, March 7, 1960, 47.

88. See, e.g., Donald Cooley, "The New Drugs That Make You Feel Better," *Cosmopolitan*, September 1956, 24–27; "The Bulging Pillbox of Tranquility," *Newsweek*, May 21, 1956, 68; "Pills for the Mind," *Time*, June 11, 1956, 54.

89. Lawrence Galton, "A New Drug Brings Relief for the Tense and Anxious," *Cosmopolitan,* August 1955, 82–83; Donald Cooley, "The New Nerve Pills and Your Health," *Cosmopolitan,* January 1956, 68–75. Donald Cooley also wrote the soft-pedaling "Story of Tranquilizers" for *Today's Health* (a physicians' waiting-room magazine), November 1960, 32–33+. The service of Galton and Cooley, regular medicine columnists and writers, for the MPIB are documented by "thumbnail sketches of writers and others whom MPIB paid for preparing backgrounders, brochures, technical memoranda etc.," in "Senate Anti-trust and Monopoly, Drugs," accession 71a 5170, record group 46, box 5, National Archives Building, Washington, DC.

90. The literature on middle-class discontents in the 1950s is legion. See chap. 2 for fuller discussion and citations.

91. Upping dosage: see, e.g., Lester Blumenthal and Marvin Fuchs, "Meprobamate: An Adjunct to Successful Management of Chronic Headache," *American Practitioner and Digest of Treatment,* July 1958, 1121–25; Frank Orland, "Use and Overuse of Tranquilizers," *JAMA,* October 10, 1959, 633–36. Quitting or reducing on their own: see, e.g., Frederick Lemere, "New Tranquilizing Drugs," *Northwest Medicine,* October 1955, 1098–1100; Walter Osinski, "Treatment of Anxiety States with Meprobamate," *Annals of the New York Academy of Sciences,* May 9, 1957, 766–71.

92. Eisenberg, "Role of Tranquilizing Drugs in Allergy," 936.

93. Refill numbers were tracked by *American Druggist'*s fortnightly prescription surveys.

94. Frederick Lemere, "New Tranquilizing Drugs," *Northwest Medicine,* October 1955, 1100.

95. Blumenthal and Fuchs, "Meprobamate: An Adjunct to Successful Management of Chronic Headache," 1123. See also Vilkin and Lomas, "Clinical Experience with Diazepam in General Psychiatric Practice," which reports the case of a patient who specifically requested therapy with Librium in order to avoid another round of electroconvulsive therapy. Such dynamics echo the arguments of Catharine Riessman, who claims that women, especially elite women, have actively participated in the "medicalization" of aspects of their lives, although their motives for doing so may have differed greatly from the medical and economic rationales of the physicians and companies with whom they collaborated. Catharine Riessman, "Women and Medicalization: A New Perspective," *Social Policy,* Summer 1983, 3–18.

96. Merritt Foster in Smith, Foster, and Rudy, "Tranquilizers and Energizers —or Neither?" 349.

97. Morgan Martin, "Pressures on Practitioners to Prescribe Tranquillizers," *Canadian Medical Association Journal,* January 16, 1960, 134.

98. T. F. Rose, "The Use and Abuse of the Tranquilizers," *Canadian Medical Association Journal*, January 15, 1958, 146. See also James Maas, "Nolle Nocere, Anxiety and Medicaments," *GP*, April 1959, 86: "Whom are we treating? The physician or the patient?"

99. Cobleigh, "The Tranquil Rise of Carter Products," 4. For comparative sales figures by company, see Standard and Poor's Industry Surveys, December 18, 1958, D12, cited in U.S. Senate, *Administered Prices in the Drug Industry*, 8910.

100. Meprotabs ad, *JAMA*, November 16, 1957, 97: "When you wish to prescribe Miltown but, for psychological reasons, not by its brand name, specify Meprotabs . . . may be prescribed as a muscle relaxant without revealing its tranquilizer action." See "Dollars and Druggists," *Forbes*, January 1, 1958, 82; Whiteside, "Getting There First with Tranquility."

Chapter Two: Listening to Miltown

1. George Beard, *American Nervousness: Its Causes and Consequences* (New York: G. P. Putnam's Sons, 1881); Gail Bederman, *Manliness and Civilization: A Cultural History of Gender and Race in the United States, 1880–1917* (Chicago: University of Chicago Press, 1995); Tom Lutz, *American Nervousness, 1903: An Anecdotal History* (Ithaca, NY: Cornell University Press, 1991); Dona Davis, "George Beard and Lydia Pinkham: Gender, Class, and Nerves in Late Nineteenth Century America," *Health Care for Women International* 10, nos. 2–3 (1989): 93–114; Francis Gosling, *Before Freud: Neurasthenia and the American Medical Community* (Urbana: University of Illinois Press, 1987), esp. 10–11, 30–32, 83–84; T. J. Jackson Lears, *No Place of Grace: Antimodernism and the Transformation of American Culture, 1880–1920* (New York: Pantheon, 1981); Barbara Sicherman, "The Uses of a Diagnosis: Doctors, Patients, and Neurasthenia," *Journal of the History of Medicine and Allied Sciences* 32, no. 1 (1977): 33–54. According to Gosling, neurasthenia among the "muscle-working" orders was described as "spinal" rather than mental and attributed to simple overwork rather than advanced character (84). See also Elaine Showalter, *Hystories: Hysterical Epidemics and Modern Culture* (New York: Columbia University Press, 1997), for a polemical critique of these illnesses in nonelite populations as well. For a broader cultural examination of attributions of psychological and emotional complexity to elites, see Nancy Schnog and Joel Pfister, eds., *Inventing the Psychological: Towards a Cultural History of Emotional Life in America* (New Haven: Yale University Press, 1997).

2. Among the voluminous literature on disease and stigma, see, e.g., Nayan Shah, *Contagious Divides: Epidemics and Race in San Francisco's Chinatown*

(Berkeley: University of California Press, 2001); Howard Markel, *Quarantine: East European Jewish Immigrants and the New York City Epidemics of 1892* (Baltimore: Johns Hopkins University Press, 1997); Naomi Rogers, *Dirt and Disease: Polio before FDR* (New Brunswick, NJ: Rutgers University Press, 1992); Alan Kraut, *Silent Travelers: Germs, Genes, and the Immigrant Menace* (Baltimore: Johns Hopkins University Press, 1994); Barron Lerner, *Contagion and Confinement: Controlling Tuberculosis along the Skid Road* (Baltimore: Johns Hopkins University Press, 1998).

3. Bederman, *Manliness and Civilization;* Lears, *No Place of Grace;* Kristin Hoganson, *Fighting for American Manhood: How Gender Politics Provoked the Spanish-American and Philippine-American Wars* (New Haven: Yale University Press, 1998); Laura Wexler, *Tender Violence: Domestic Visions in an Age of U.S. Imperialism* (Chapel Hill: University of North Carolina Press, 2000); Lutz, *American Nervousness, 1903;* John Higham, "The Reorientation of American Culture in the 1890s," in *The Culture of Consumption: Critical Essays in American History, 1880–1980,* ed. Richard W. Fox and T. J. Jackson Lears (New York: Pantheon, 1983).

4. Anne Lane, ed., *The Charlotte Perkins Gilman Reader* (Charlottesville: University of Virginia Press, 1999). See also, e.g., Susan Curtis, *A Consuming Faith: The Social Gospel and Modern American Culture* (Baltimore: Johns Hopkins University Press, 1991).

5. Ellen Herman, *The Romance of American Psychology: Political Culture in the Age of Experts* (Berkeley: University of California Press, 1995); Gerald Grob, *From Asylum to Community: Mental Health Policy in Modern America* (Princeton: Princeton University Press, 1991); Nathan Hale Jr., *The Rise and Crisis of Psychoanalysis in the United States: Freud and the Americans, 1917–1985* (New York: Oxford University Press, 1995); Eva Moskowitz, *In Therapy We Trust: America's Obsession with Self-Fulfillment* (Baltimore: Johns Hopkins University Press, 2001); Elizabeth Lunbeck, *The Psychiatric Persuasion: Knowledge, Gender, and Power in Modern America* (Princeton: Princeton University Press, 1994).

6. The New York City study was Leo Srole et al., *Mental Health in the Metropolis* (New York: McGraw-Hill, 1962). See also Benjamin Pasamanick, ed., *Epidemiology of Mental Disorder: A Symposium Organized by the American Psychiatric Association* (Washington, DC: American Association for the Advancement of Science, 1959); Arnold Rose et al., *Mental Health and Mental Disorder: A Sociological Approach* (New York: Norton, 1955); Paul Hoch and Joseph Zubin, eds., *Comparative Epidemiology of the Mental Disorders* (New York: Grune and Stratton, 1961); Derek Phillips, "The 'True Prevalence' of Mental Illness in a New England State," *Community Mental Health Journal,* Spring 1966, 35–40. For a broader look at these epidemiological studies in historical context, see Gerald

Grob, "The Origins of American Psychiatric Epidemiology," *American Journal of Public Health*, March 1985, 229–36.

7. August Hollingshead and Frederick Redlich, *Social Class and Mental Illness: A Community Study* (New York: John Wiley and Sons, 1952), 171–93, 222–33, 235, 239–40.

8. Ibid., 339–40; Jack Ewalt and Dana Farnsworth, *Textbook of Psychiatry* (New York: McGraw-Hill, 1963), 98.

9. Rollo May, *The Meaning of Anxiety* (New York: Ronald, 1950), 344–45. The federal Joint Commission on Mental Illness and Health followed a similar line in its mammoth survey *Americans View Their Mental Health*, reporting that men in high-status jobs suffered "greater worry and distress" than those in less desirable positions because of their higher aspirations. See Gerald Gurin, Joseph Veroff, and Sheila Feld, *Americans View Their Mental Health: A Nationwide Survey*, Joint Commission on Mental Illness and Health, Monograph series 4 (New York: Basic Books. 1960), xvii–xviii. An interesting example of such distinctions can also be found in psychoanalyst Frieda Fromm-Reichmann's distinction between fear and anxiety: "fear is a useful, rational kind of fright elicited by realistic *external* dangers," while anxiety stems from "the *inner* danger of unacceptable thoughts, feelings, wishes, or drives." True anxiety (an illness) emerged from one's inner life; simple animal fear was a direct response to external stimuli. See Frieda Fromm-Reichmann, "Psychiatric Aspects of Anxiety," in *Identity and Anxiety: Survival of the Person in Mass Society*, ed. Maurice Stein, Arthur Vidich, and David Manning White (Glencoe, IL: Free Press, 1960), 130–31.

10. "Tension and the Nerves of the Nation . . . Psychiatry Eyes the Breaking Point," *Newsweek*, March 5, 1956, 54–58; Donald G. Cooley, "The Story of Tranquilizers," *Today's Health*, November 1960, 58–59; Margaret Mead, "One Vote for This Age of Anxiety," *New York Times Magazine*, May 20, 1956, 18, 56–58; "The Anatomy of *Angst*," *Time*, March 31, 1961, 44–51.

11. For a broader argument on the attribution of complex interiority to elites, see Pfister, introduction to *Inventing the Psychological*, ed. Schnog and Pfister, xx.

12. K. A. Courdileone, "Politics in an Age of Anxiety," *Journal of American History*, September 2000, 515–45, and *Manhood and American Political Culture in the Cold War* (New York: Routledge, 2005); Jonathan Metzl, *Prozac on the Couch: Prescribing Gender in the Era of Wonder Drugs* (Durham, NC: Duke University Press, 2003). For a critique of present-day scholars' use of the "gender crisis" paradigm, see Bryce Traister, "Academic Viagra: The Rise of American Masculinity Studies," *American Quarterly* 52 (2000): 274–304.

13. David Riesman, *The Lonely Crowd: A Study of the Changing American*

Character (New Haven: Yale University Press, 1950), 25–26; Norman Mailer, "The White Negro," in *Advertisements for Myself* (New York: Putnam, 1959). See also Editors of Look, *The Decline of the American Male* (New York: Random House, 1958).

14. Courdileone, *Manhood and American Political Culture in the Cold War*, 516, citing Arthur Schlesinger's *The Vital Center: The Politics of Freedom* (Boston: Houghton Mifflin, 1962) and "The Crisis of American Masculinity" in *The Politics of Hope* (Boston: Houghton Mifflin, 1962), 237–46, originally published in *Esquire*.

15. Robert Dean, *Imperial Brotherhood: Gender and the Making of Cold War Foreign Policy* (Amherst: University of Massachusetts Press, 2001); Andrea Friedman, "The Smearing of Joe McCarthy: The Lavender Scare, Gossip, and Cold War Politics," *American Quarterly* 57 (December 2005): 1105–29; John D'Emilio, "The Homosexual Menace: The Politics of Sexuality in Cold War America," in *Making Trouble: Essays on Gay History, Politics, and the University* (New York: Routledge, 1992); David Johnson, *The Lavender Scare: The Cold War Persecution of Gays and Lesbians in the Federal Government* (Chicago: University of Chicago Press, 2004).

16. Ferdinand Lundberg and Marynia Farnham, *Modern Woman: The Lost Sex* (New York: Harper and Brothers, 1947), 10. See also Philip Wylie, *Generation of Vipers* (New York: Rinehart, 1942).

17. Stephanie Coontz, *The Way We Never Were: American Families and the Nostalgia Trap* (New York: Basic Books, 1992); Metzl, *Prozac on the Couch*, 77–98; Elaine Tyler May, *Homeward Bound: American Families in the Cold War Era* (New York: Basic Books, 1988); Mari Jo Buhle, *Feminism and Its Discontents: A Century of Struggle with Psychoanalysis* (Cambridge: Harvard University Press, 2000).

18. For similar logic about alcoholism, see Lori Rotskoff, *Love on the Rocks: Men, Women, and Alcohol in Post–World War II America* (Durham: University of North Carolina Press, 2001).

19. Gerhard Haugen, Herman Dickel, and Henry Dixon, *A Therapy for Anxiety and Tension Reactions* (New York: Macmillan, 1958), 12–13; Herman Dickel, James Wood, and Henry Dixon, "Electromyographic Studies on Meprobamate and the Working, Anxious Patient," *Annals of the New York Academy of Sciences*, May 9, 1957, 781.

20. Conference reprinted in the *Journal of Neuropsychiatry*. Rickels's presentation was "Important and Relevant Aspects of Tranquilizer Therapy," *Journal of Neuropsychiatry*, 1964, 442–45. See also Karl Rickels, "The Use of Psychotherapy with Drugs in the Treatment of Anxiety," *Psychosomatic Medicine*, 1964, 114.

21. This is a remarkable omission given the media attention to *Brown v. Board of Education,* which turned in certain key respects on studies about the psychological harm segregation inflicted on African Americans. See Moskowitz, *In Therapy We Trust.* One fascinating counterexample can be found in a Tennessee physician's suggestion to use tranquilizers to "inoculate" troubled neighborhoods to forestall social unrest from delinquents and criminals. This attitude was so far from the mainstream, however, that it was the exception that proves the rule. See P. Sottong, "Public Health Aspects of the Tranquilizing Drugs," *Journal of the Tennessee State Medical Association,* January 1958, 12–17.

22. "Driven himself" and "business executive" from Frederick Lemere, "New Tranquilizing Drugs," *Northwest Medicine,* October 1955, 1100; lawyers and overtaxed housewives from Joseph Borrus, "Meprobamate in Psychiatric Disorders," *Medical Clinics of North America,* 1957, 327, 330, and Lawrence Linn, *Journal of the Mount Sinai Hospital,* 1958, 141–42; psychiatry textbook is Frederic Flach and Peter Regan, *Chemotherapy in Emotional Disorders: The Psychotherapeutic Use of Somatic Treatments* (New York: McGraw-Hill, 1960), 132–39.

23. The advertisement ran prominently on the front inside cover of *JAMA.* It featured two pictures, one of a black man with a caption explaining that his "tensions are part of everyday life, and can be handled without a doctor." The caption of the second picture, of a white woman, reads, "Some kinds are distressing to patients—they need your reassurance and support." Valium advertisement *JAMA,* October 6, 1978, front cover. In general, racial exclusivity was not unusual in the advertising industry, lay or medical, but neither was it an absolute rule: according to a 1974 study of sixty medical journals, African-American models had found their way into a minority of ads for some nonpsychotropic categories of drugs by the early 1970s. Mickey C. Smith, "Where Are the Blacks in Prescription Drug Advertising?" *Medical Marketing and Media,* May 1977, 47–49.

24. Dexamyl ad, *American Journal of Psychiatry,* 1959, xxvi–xxvii.

25. *American Druggist,* fortnightly prescription surveys, 1955–1964. For health insurance, see, e.g., Odin Anderson, Patricia Collette, and Jacob Feldman, *Changes in Family Medical Care Expenditures and Voluntary Health Insurance: A Five-Year Resurvey* (Cambridge: Harvard University Press, 1963), 72–73, 39–40; Herman and Anne Somers, *Doctors, Patients, and Health Insurance: The Organization and Financing of Medical Care* (Washington, DC: Brookings Institution, 1962), 389–91; Jonathan Engel, *Poor Peoples' Medicine: Medicaid and American Charity Care since 1965* (Durham, NC: Duke University Press, 2006).

26. The discrepancy by economic class was not large, however, and it did not appear in all surveys. For studies that tracked use by race and class, see Hugh

Parry, "Use of Psychotropic Drugs by U.S. Adults," *Public Health Reports,* October 1968, 799–810; Carl Chambers, "An Assessment of Drug Use in the General Population," in *Drug Use and Social Policy: An AMS Anthology,* ed. Jackwell Susman (New York: AMS, 1972), 50–123; *Drug Use in America: Problem in Perspective: Second Report of the National Commission on Marihuana and Drug Abuse* (Washington, DC: GPO, 1973), 41–119; Jonathan Davidson, David Raft, B. Frank Lewis, and Margaret Gebhardt, "Psychotropic Drugs on General Medical and Surgical Wards of a Teaching Hospital," *AMA Archives of General Psychiatry,* April 1975, 507–11; George Warheit, Sandra Arey, and Edith Swanson, "Patterns of Drug Use: An Epidemiologic Overview," *Journal of Drug Issues,* Summer 1976, 223–37; E. H. Uhlenhuth, Mitchell Balter, and Ronald Lipman, "Minor Tranquilizers: Clinical Correlates of Use in an Urban Population," *AMA Archives of General Psychiatry,* May 1978, 650–55; Richard Tessler, Randall Stokes, and Marianne Pietras, "Consumer Response to Valium," *Drug Therapy,* February 1978, 178–83. Dean Manheimer, Glen Mellinger, and Mitchell Balter, "Psychotherapeutic Drugs: Use among Adults in California," *California Medicine,* December 1968, 445–51, did not show much discrepancy by race or by class but, tellingly, also did not distinguish between prescription drugs and over-the-counter medications.

27. Moskowitz, *In Therapy We Trust;* Herman, *Romance of American Psychology;* Grob, *From Asylum to Community;* Rotskoff, *Love on the Rocks.*

28. See, e.g., Milton Moskowitz, "Librium Becomes No. 1 Tranquilizer, Though Outspent by Rival Advertiser," *Advertising Age,* September 12, 1960, 3, 66–67. See also Moskowitz, "Librium: A Marketing Case History," *Drug and Cosmetic Industry* 87, 1960, 460–61, 566–67. For examples from other financial media, see chap. 1, note 21.

29. "Don't-Give-a-Damn Pills," *Time,* February 27, 1956, 98; "Happiness by Prescription," *Time,* March 11, 1957, 59. See also, e.g., Alek Rozental, "The Strange Ethics of the Pharmaceutical Industry," *Harper's,* May 1960, 78–79; Barbara Youncker, "Pills and You," *New York Post,* September 13, 1962; Robert Goldman, "Instant Happiness," *Ladies Home Journal,* October 1963, 67–71.

30. Eric Hodgins, "The Search Has Only Started," *Life,* October 22, 1956, 139. The cartoon originally ran in the *New Yorker.*

31. "Pills vs. Worry—How Goes the Frantic Quest for Calm in Frantic Lives?" *Newsweek,* May 21, 1956, 68, photo 70.

32. "'Wonder Drugs' and Mental Disorders," *Consumer Reports,* August 1955, 386–89; "Unsettling Facts about Tranquilizers," *Consumer Reports,* January 1958, 4; "The High Cost of Rx Drugs," *Consumer Reports,* November 1958, 597–99.

33. "Big Pill Bill to Swallow: The Wonder-Drug Makers Get Handsome Prof-

its from Their Captive Consumers," *Life*, February 15, 1960, 97–103. Abbott Laboratories responded by placing a full-page "Paid Editorial" rebutting the charges. "A Paid Editorial: Report to the Nation's Doctors on 'A Big Pill Bill to Swallow,'" *Life*, March 23, 1960, 121.

34. Francis Bello, "The Tranquilizer Question," *Fortune*, May 1957, 162–63.

35. "Tranquil Extinction," *Business Week*, October 27, 1956, 204.

36. Dr. James H. Wall quoted in "New Avenues into Sick Minds," *Life*, October 22, 1956, 140. See also, e.g., Arthur Gordon, "Happiness Doesn't Come in Pills," *Reader's Digest*, January 1957, 60–62 (condensed from *Woman's Day*, January 1957).

37. Quotations, in order, from Roland Berg, "Unhappy Facts about Happy Pills," *Look*, July 24, 1956, 92; Albert Deutsch, "What Anxiety Can Do for You," *Woman's Home Companion*, November 1956, 55 (also reprinted in *Catholic Digest* as "The Danger in 'Happy' Pills," November 1956, 11–15); "What You Ought to Know about Tranquilizers," *U.S. News and World Report*, June 21, 1957, 64; "Domestic Tranquility," *New Republic*, June 24, 1957, 5. For similar examples, see, e.g., Marie Nyswander, "The Pill and You," *Vogue*, June 1957, 140; Edward Podolsky, "The Facts on 'Happy Pills,'" *Catholic Home Journal*, October 1957, 30.

38. "New Avenues into Sick Minds," *Life*, October 22, 1956, 140–41; "Tranquil Extinction," *Business Week*, October 27, 1956, 204; Jess Raley, "That Wonderful Frustrated Feeling," *American Mercury*, July 1957, 24 ("The most wonderful thing about the tranquilizing drugs is that they didn't happen sooner," the article continued, and now that they have been, will not civilization's "pace slacken, cease to move, and begin to slip backwards?"). See also, e.g., Ardis Whitman, "Learn to Live with Your Worries," *Reader's Digest*, February 1956, 113 (condensed from *Journal of Lifetime Living*, February 1956); "No Peace for Tranquilizers," *Business Week*, September 1, 1956, 32; Bello, "Tranquilizer Question," 165; "Pills to Lull Children," *Newsweek*, November 10, 1958, 82; T. F. James, "The Modern Mind," *Cosmopolitan*, April 1958, 22–29; Sir Geoffrey Vickers quoted in "Excess in the Cult of Sanity," *America*, August 28, 1954, 510.

39. Christopher Shaw and Philip Felts, "Treacherous Tranquilizers," *American Journal of the Medical Sciences*, February 1959, 142; Herman Dickel and Henry Dixon, "Inherent Dangers in Use of Tranquilizing Drugs in Anxiety States," *JAMA*, February 9, 1957, 425–26. See also, e.g., Cyril Conway Jr., "Tranquilizers: Their Use and Abuse," *Transactions and Studies of the College of Physicians of Philadelphia*, July 1960, 38.

40. T. F. Rose, "Use and Abuse of the Tranquilizers," *Canadian Medical Association Journal*, January 15, 1958, 148; Elizabeth Kahler, "Sedatives, Tranquilizers—Then What?" *Journal of the American Medical Women's Association*, July 1957, 225–26; Harry Beckman, "Introductory Remarks," in"Meprobamate and

Other Agents Used in Mental Disturbances," ed. Frank Berger, *Annals of the New York Academy of Sciences*, May 9, 1957, 673–74.

41. Aldous Huxley, "Drugs That Shape Men's Minds," *Saturday Evening Post*, October 18, 1958, 111. Huxley would probably have been appalled to discover that Miltown's manufacturer, Carter Products, already had a drug called Soma on the market (it was a painkiller). Other science fiction writers had imagined tranquilizers before they were actually discovered. See, e.g., John MacDonald's 1949 "Trojan Horse Laugh," *Astounding Science Fiction* 43, no. 6:73–111, in which a well-advertised and highly popular treatment to synchronize people's monthly cycles of joy and sadness (thus producing social harmony) turns out to be a Russian plot. Once a critical mass of happiness or depression is achieved, orgies or riots tear apart American cities and the Red Army invades with impunity. See also Arthur Barnes, "Emotion Solution," *Wonder Stories* 7, no. 8:955–63, in which a scientist invents a drug to put humans on mammalian estrus cycles, unintentionally leading to total female domination of periodically sex-obsessed men. Robert Silverberg's *Drug Themes in Science Fiction* (Rockville, MD: National Institute on Drug Abuse, 1974) provides a comprehensive annotated bibliography from 1900.

42. [Editors], "A Decade of Tranquilizers," *Journal of the Medical Society of New Jersey*, July 1964, 245–46.

43. David Cowen, "Those Pretty Little Pills," *Nation*, April 16, 1960, 338. See also "Happiness Pills Are No Answer," *Christian Century*, September 12, 1958, 1044; "Happiness by Prescription," *Time*, March 11, 1957, 59; Thomas Whiteside, "Getting There First with Tranquility," *New Yorker*, May 3, 1958, 127.

44. Quoted in Dean, *Imperial Brotherhood*, 172.

45. Striatran ad, *American Journal of Psychiatry*, September 1960.

46. Librium ad, *American Journal of Psychiatry*, January 1962.

47. Valium ad, *JAMA*, May 31, 1965. See Metzl, *Prozac on the Couch*, for a different argument about the drugs helping men cope with the threat from overbearing or "Momist" women.

48. Valium ad, *American Journal of Psychiatry*, August 1965; Librium ad, *JAMA*, February 15, 1965.

49. Both Carter Products and Roche Pharmaceuticals came out with combination tranquilizers and stomach-soothers (Milpath and Librax, respectively) that appear to have been efforts to access the male market from a side entrance. For some of many examples, see Valium, the "Somatic Mask" series (e.g., *JAMA* January 4 and May 31, 1965) and "somatic symptoms" series (e.g., *JAMA*, July 22 and September 2, 1968); Valium ad, *JAMA*, April 21, 1969; Librium ad, *JAMA*, July 3, 1967, February 12, 1968, and August 4, 1969; Librax ad, *JAMA*, July 10 and September 11, 1967; Milpath ad, *JAMA*, July 17, 1967; Miltrate ad, *JAMA*,

January 8, 1968. For images, see *JAMA*, February 19, 1968 (Milpath); *JAMA*, February 12, 1968 (Librium).

50. See, e.g., Valium ad, *JAMA*, May 11, 1970. Another Valium ad featured an image of a working man getting a "skeletal muscle spasm" from lifting a heavy "rush" package for mailing. See *JAMA*, January 20, 1975.

51. This ironically mirrored Freudian dynamic models of the mind, in which anxiety and its countless related ills were caused by psychic conflicts between basic animal drives and the restrictive codes of civilization.

52. Librium ads, *American Journal of Psychiatry*, May 1960. See also Valium ad, *JAMA*, April 7, 1969 ("Animal Pharmacology: Calming and Taming the Monkey"); Librium ad, *JAMA*, October 23, 1972 (in which "radio-controlled ESB pinpoints action of Librium on selected brain areas of rhesus monkeys"); Librium ad, *JAMA*, March 5, 1973 ("How radiostimulated aggression in freely moving monkeys sheds new light on mechanisms of their behavior").

53. Cooley, "Story of Tranquilizers."

54. Librium ad, *JAMA*, September 15, 1969. See also Etrafon ad, *JAMA*, March 11, 1968, 75, which shows the image of a primitive headhunter next to a modern businessman.

55. Valium ad, *JAMA*, January 8, 1968, 231–39. See also ad for Librax (tranquilizer plus antacid for ulcers) in *JAMA*, September 27, 1971.

56. Desmond Morris, *The Naked Ape: A Zoologist's Study of the Human Animal* (New York: McGraw-Hill, 1967). For a historical overview of the new "human zoology," see Carl Degler, *In Search of Human Nature: The Decline and Revival of Darwinism in American Social Thought* (New York: Oxford University Press, 1991).

57. Dean, *Imperial Brotherhood*.

58. Parry, "Use of Psychotropic Drugs," 802, 808. See also Manheimer, Mellinger, and Balter, "Psychotherapeutic Drugs," 448; Mitchell Balter and Jerome Levine, "The Nature and Extent of Psychotropic Drug Usage in the United States," *Psychopharmacology Bulletin*, October 1969, 13; Glen Mellinger, Mitchell Balter, and Dean Manheimer, "Patterns of Psychotherapeutic Drug Use among Adults in San Francisco," *AMA Archives of General Psychiatry*, November 1971, 393–94; Mitchell Balter and Jerome Levine, "Character and Extent of Psychotherapeutic Drug Usage in the United States," in *Psychiatry: Proceedings of the V World Congress of Psychiatry, Mexico, D.F., 25 November–4 December, 1971*, ed. Ramón de la Fuente and Maxwell Weisman (New York: American Elsevier, 1973), 87–88; Hugh Parry et al., "National Patterns of Psychotherapeutic Drug Use," *AMA Archives of General Psychiatry*, June 1973, 769. Then as now, women used the medical system more than men did, and this may have accounted for some of the unevenness in drug prescribing, but it could not account for it all. A

mid-1970s study of benzodiazepine use at the University of Rochester's Family Medical Center, for example, found 72% of all benzodiazepine recipients to be women. The study further revealed that 9.3% of all female patients received a prescription for benzodiazepine, as opposed to 4.8% of the male patients. Of patients who received at least one prescription from the center, 13.1% of women received benzodiazepine, whereas only 8.6% of men did so. To put the same figures slightly differently, 7.7% of women's prescriptions were for benzodiazepines, while only 6.1% of men's were. Jeffrey Hasday and Fred Karch, "Benzodiazepine Prescribing in a Family Medicine Center," *JAMA*, September 18, 1981, 1321–25. Nor was this sex disparity typical for drugs in general. National figures from the 1970s indicated that 60% of all prescriptions were given to women, but 60% of all patients were in fact women; both men and women had a prescription-per-doctor-visit ratio of 1.09:1. Carlene Baum et al., "Drug Use in the United States in 1981," *JAMA*, March 9, 1984, 1293–97.

59. May, *Homeward Bound;* Buhle, *Feminism and Its Discontents.* As Joanne Meyerowitz noted in her edited volume *Not June Cleaver: Women and Gender in Postwar America, 1945–1960* (Philadelphia: Temple University Press, 1994), such cult-of-domesticity messages may have been prominent but were hardly the only ones circulating about women at the time.

60. Meprospan ad, *JAMA*, July 16, 1960.

61. Dexamyl ads, *American Journal of Psychiatry,* August 1965 and January 1964. Similar example in Ritalin ads: a woman's face trapped in a huge pile of dishes, a man's face trapped in a huge pile of in-box papers. *American Journal of Psychiatry,* February and May 1965.

62. Sinequan ad, *JAMA*, October 12 and September 7, 1970. A Serpasil (antipsychotic) ad pictured a housewife screaming at her son, dressed as a cowboy, who has disturbed her as she tries to read the paper and drink a cup of coffee on a break from vacuuming. *JAMA*, April 21, 1956, 9. Dexamyl ad, *JAMA*, June 15, 1957, 30. In 1972, an ad for the antidepressant Triavil instructed physicians that it might be depression if a patient "finds it 'almost impossible to start housework every morning.'" Triavil ad, *JAMA*, February 21, 1972.

63. Vivactil ad, *JAMA*, April 13, 1970.

64. Daniel Horowitz, "Rethinking Betty Friedan and *The Feminine Mystique*: Labor Union Radicalism and Feminism in Cold War America," *American Quarterly* 48 (1996): 1–42.

65. Betty Friedan, *The Feminine Mystique* (New York: Norton, 1963), 293 ("mild, undiagnosable"), 20–21 ("empty somehow," "tired feeling," "strange feeling"), 235 ("dull and flat").

66. Ibid., 26 ("material advantages"), 235 ("to be envied, " "uncommonly smart"), 294 ("below the level").

67. Ibid., 20 ("blotted"), 31 ("many suburban housewives"), 310 ("You wake up").

68. Ibid., 314.

69. Miltown ad, *JAMA*, January 22, 1968; Serax ad, *Medical Aspects of Human Sexuality*, December 1969, reprinted in U.S. Senate, Select Committee on Small Business, Subcommittee on Monopoly, *Advertising of Proprietary Medicines, Part 2: Mood Drugs (Sedatives, Tranquilizers, and Stimulants)*, 92nd Cong., 1st sess., July 21, 22, 23, September 22, 1971 (Washington, DC: GPO, 1971), 868–69.

Chapter Three: Wonder Drugs and Drug Wars

1. Frederick Lemere, ""New Tranquilizing Drugs," *Northwest Medicine*, October 1955, 1098.

2. Frederick Lemere, "Drug Habituation," *JAMA*, April 21, 1956, 1431, and "Habit Forming Properties of Meprobamate," *AMA Archives of Neurology and Psychiatry*, August 1956, 205–6.

3. For a book-length assessment of these distinctions, see Richard DeGrandpre, *The Cult of Pharmacology: How America Became the World's Most Troubled Drug Culture* (Durham, NC: Duke University Press, 2006).

4. David Courtwright, *Dark Paradise: A History of Opiate Addiction in America* (Cambridge: Harvard University Press, 2001); Caroline Jean Acker, *Creating the American Junkie: Addiction Research in the Classic Era of Narcotic Control* (Baltimore: Johns Hopkins University Press, 2002), 18–42, and "From All Purpose Anodyne to Marker of Deviance: Physicians' Attitudes towards Opiates in the US from 1890 to 1940," in *Drugs and Narcotics in History*, ed. Roy Porter and Mikalàs Teich (New York: Cambridge University Press, 1995), 114–32; Joseph Spillane, *Cocaine: From Medical Marvel to Modern Menace in the United States, 1884 to 1920* (Baltimore: Johns Hopkins University Press, 2002); David Musto, *The American Disease: Origins of Narcotic Control*, 3rd ed. (New York: Oxford University Press, 1999); Nayan Shah, *Contagious Divides: Epidemics and Race in San Francisco's Chinatown* (Berkeley: University of California Press, 2001).

5. Acker, *Creating the American Junkie*, 18–42; Courtwright, *Dark Paradise*; DeGrandpre, *Cult of Pharmacology*; Spillane, *Cocaine*. Spillane notes, however, that antivice campaigns even before the 1920s tended to result in arrests of urban cocaine users, even though they were not targeted specifically for using cocaine. See also Caroline Jean Acker, "Portrait of an Addicted Family: Dynamics of Opiate Addiction in the Early Twentieth Century," in *Altering American Consciousness: The History of Alcohol and Drug Use in the United States, 1800–*

2000, ed. Sarah Tracy and Caroline Jean Acker (Amherst: University of Massachusetts Press, 2004), 165–81, who suggests that the difference between pre– and post–Harrison Act addicts might not be as stark as is commonly believed.

6. Musto, *American Disease;* Acker, *Creating the American Junkie;* Spillane, *Cocaine;* Jim Baumohl, "Maintaining Orthodoxy: The Depression-Era Struggle over Morphine Maintenance in California," in *Altering American Consciousness,* ed. Tracy and Acker, 225–66.

7. Lawrence Kolb, "Pleasure and Deterioration from Narcotic Addiction," *Mental Hygiene* 19 (1925): 723, quoted in Musto, *American Disease,* 84. See also Acker, *Creating the American Junkie;* William White, *Slaying the Dragon: The History of Addiction Treatment and Recovery in America* (Bloomington, IL: Chestnut Health Systems/Lighthouse Institute, 1998); Timothy Hickman, "The Double Meaning of Addiction: Habitual Narcotic Use and the Logic of Professionalizing Medical Authority in the United States, 1900–1920," in *Altering American Consciousness,* ed. Tracy and Acker, 182–202.

8. For Los Angeles antimarijuana campaigns, see Curtis Marez, *Drug Wars: The Political Economy of Narcotics* (Minneapolis: University of Minnesota Press, 2004). For a similar racialized logic regarding "opium dens," see Shah, *Contagious Divides.* See also Musto, *American Disease,* and Acker, *Creating the American Junkie.*

9. Musto, *American Disease,* 219; Marez, *Drug Wars;* DeGrandpre, *Cult of Pharmacology.* For popular culture treatments of marijuana, see, e.g., Harry Anslinger, "Marijuana: Assassin of Youth," *American Magazine,* July 1937, 18–19+, and the movie *Reefer Madness* (also referred to as *Tell Your Children,* 1936).

10. Acker, *Creating the American Junkie,* 125–55; Spillane, *Cocaine.*

11. See, e.g., *American Druggist* prescription surveys from the 1950s; Charles O. Jackson, "Before the Drug Culture: Barbiturate/Amphetamine Abuse in American Society," *Clio Medica* 11, no. 1 (1976): 47–58; Acker, *Creating the American Junkie,* 125–55.

12. "Federal Food, Drug, and Cosmetic Act," Public Law 75–717 (52 *U.S. Stat* 1040), 25 June 1938, sections 502, 503, and 505; "An Act to Amend Sections 303(c) 503(b) of the Federal Food, Drug, and Cosmetic Act," 82nd Cong., 1st sess., October 26, 1951, 65 *U.S. Stat* 648 (Durham-Humphrey Amendment). For quotation from Section 502(d) of the 1938 act ("pleasurable stimulation or escape"), see *F-D-C Reports,* June 13, 1955, 14–17. See also Jackson, "Before the Drug Culture"; James Harvey Young, "Federal Drug and Narcotic Legislation," *Pharmacy in History* 37 (1995): 59–67; John P. Swann, "FDA and the Practice of Pharmacy: Prescription Drug Regulation before the Durham-Humphrey Amendment of 1951," *Pharmacy in History* 36 (1994): 55–70.

13. Acker, *Creating the American Junkie,* 75–77.

14. Ibid., 84–87.

15. Testimony of Dr. Kenneth W. Chapman, consultant on narcotic addiction, National Institute of Mental Health, U.S. Public Health Service, in U.S. House, Committee on Ways and Means, *Traffic in, and Control of, Narcotics, Barbiturates, and Amphetamines,* 84th Cong., 1st sess., October 13, 1955 (Washington, DC: GPO, 1956), 61.

16. See Harris Isbell's testimony and publications, ibid., 295–421. Also White, *Slaying the Dragon,* 126. In 1959 Isbell defined addiction as a "condition brought about by repeated administration of a drug which results in an altered physiological state," a definition that prompted him to declare marijuana, cocaine, and amphetamines nonaddictive but barbiturates and tranquilizers addictive. See Isbell, "Addiction to Hypnotic and Sedative Drugs," *Association of Food and Drug Officials of the U.S.* 23 (1959): 35.

17. Carl Essig and John Ainslie, "Addiction to Meprobamate," *JAMA,* July 20, 1957, 1382; Carl Essig, "Withdrawal Convulsions in Dogs Following Chronic Meprobamate Intoxication," *AMA Archives of Neurology and Psychiatry* 80 (October 1958): 414, 417.

18. Nathan Eddy, "'Addiction-Producing' versus 'Habit-Forming' Drugs," *JAMA,* April 27, 1957, 1622–23.

19. U.N. Commission on Narcotic Drugs, report of the 12th sess., April 29–May 31, 1957, Official Records, Economic and Social Council, 24th sess., suppl. no. 10, 49; cited in U.S. Senate, Committee on Government Operations, Subcommittee on Reorganization and International Organizations, *Interagency Coordination in Drug Research and Regulation* (Washington, DC: GPO, 1964), 1414–15.

20. Narcotic Control Act of 1956, Public Law 728, 84th Cong., 2nd sess. See Musto, *American Disease,* 231–32.

21. U.S. House, *Traffic in, and Control of, Narcotics, Barbiturates, and Amphetamines,* 28, 42, 45 (quotations from assistant surgeon general and NIMH consultant), 290–94 (Lexington movie transcript), 295–421 (Isbell's testimony), 546–53 (Trichter's testimony).

22. Ibid., 253, 260 (AMA opposition), 266–68 (American Pharmaceutical Association, American Pharmaceutical Manufacturers Association, National Wholesale Druggists' Association, etc.), 1412 (American Dental Association). These dramas can be usefully tracked in *F-D-C Reports "Pink Sheet,"* a trade journal following government developments of relevance to food, drug, and cosmetic manufacturers. See, e.g., "House Subcommittee Asks about Tranquilizer Addiction, Label Claims," *F-D-C Reports,* March 25, 1957, 10–11; "Barb-Amphetamine 'Addiction' Cited by FDA to Support Boggs Bill: 'Illegal Possession' Compromise Offered House Subcommittee," *F-D-C Reports,* June 10, 1957, 5–6; "American

Psychiatric Association's Kline Hits Unfavorable Tranquilizer Publicity at Senate Hearing: National Institutes of Health Doctor Urges Nonrefillable Prescriptions," *F-D-C Reports*, June 10, 1957, 10–11.

23. Exhibit 193, "Psychopharmacologicals: Excerpts of Views during the Eight-Year Period, 1955–1963, on Habituation and Other Aspects," in U.S. Senate, *Interagency Coordination in Drug Research and Regulation*, 1421–22; originally cited in U.S. Senate, Committee on the Judiciary, Subcommittee on Antitrust and Monopoly, *Administered Prices in the Drug Industry, Part 22* (Washington, DC: GPO, 1959–61), 12031–32.

24. U.S. House, *Traffic in, and Control of, Narcotics, Barbiturates, and Amphetamines*, 412–14.

25. Ibid., 44.

26. Ibid., 260–61.

27. Ibid., 266–68.

28. Ibid., 193; Anslinger interview with David Musto, cited in Musto, *American Disease*, 213.

29. See Nat Hentoff, *A Doctor among the Addicts* (New York: Rand McNally, 1967), and Marie Nyswander, *The Drug Addict as a Patient* (New York: Grune and Stratton, 1956). Quotation from Nyswander, 13. White, *Slaying the Dragon*, 126.

30. *Drug Addiction: Crime or Disease? Interim and Final Reports of the Joint Committee of the American Bar Association and the American Medical Association on Narcotic Drugs* (Bloomington: Indiana University Press, 1961).

31. *Robinson v. California*, 370 U.S. 660. See Musto, *American Disease*, for a discussion of this shift to "drug toleration," or John Burnham, *Bad Habits: Drinking, Smoking, Taking Drugs, Gambling, Sexual Misbehavior, and Swearing in American History* (New York: New York University Press, 1993), for a countervailing view.

32. Lori Rotskoff, *Love on the Rocks: Men, Women, and Alcohol in Post–World War II America* (Durham: University of North Carolina Press, 2001); Ron Roizen, "How Does the Nation's 'Alcohol Problem' Change from Era to Era?" in *Altering American Consciousness*, ed. Tracy and Acker, 61–87; Michelle McClellan, "'Lady Tipplers': Gendering the Modern Alcoholism Paradigm, 1933–1960," ibid., 267–97; Rotskoff, "Sober Husbands and Supportive Wives: Marital Dramas of Alcoholism in Post–World War II America," ibid., 298–326. For AA's rejection of tranquilizers, see Alcoholics Anonymous, *Tranquilizers, Sedatives, and the Alcoholic: Six A.A. Case Histories* (New York: AA World Services, 1959).

33. President's Advisory Commission on Narcotic and Drug Abuse, *Final Report* (Washington, DC: GPO, 1963), 1–7. Not surprisingly, along with the

recognition of "respectable" addicts came a more conciliatory approach to dealing with addicts: "The definition of legitimate medical use of narcotic drugs and legitimate medical treatment of a narcotic addict are primarily to be determined by the medical profession" (8).

34. "Addiction-Producing Drugs," *JAMA*, July 6, 1957, 1150.

35. Eddy, "'Addiction-Producing' versus 'Habit-Forming' Drugs," 1622–23.

36. Ibid.

37. "Addiction-Producing Drugs," *JAMA*, July 6, 1957, 1150.

38. Eddy, "Addiction-Producing versus 'Habit-Forming' Drugs," 1622–23; "Addiction-Producing Drugs," *JAMA*, July 6, 1957, 1150.

39. Nathan Eddy et al., "Drug Dependence: Its Significance and Characteristics," *Bulletin of the World Health Organization*, 1965, 722, 723, 726. See also Carl Essig, "New Sedative Drugs That Can Cause States of Intoxication and Dependence of Barbiturate Type," *JAMA*, May 23, 1966, 126–29.

40. Eddy et al., "Drug Dependence," 723.

41. Detailed (if partisan) coverage of these legislative developments can be found in the drug trade journal *F-D-C Reports*. See, e.g., "White House Commission Asks for Inclusion of Tranquilizers along with Amphetamines and Barbiturates in Control Law," April 8, 1963, 16; "Psychotoxic Drug Definition," October 14, 1963, 5; "Meprobamate, Librium, Thorazine, Other Tranquilizers Should Get Same Strict Controls Advocated for Amphetamines and Barbiturates: White House Commission," January 27, 1964, 6–11.

42. President's Advisory Commission, *Final Report*, 4, 2, 7, 1. "Psychotoxic" appeared in the 1965 WHO guidelines.

43. U.S. Senate, Committee on Labor and Public Welfare, Subcommittee on Health, *Control of Psychotoxic Drugs*, 88th Cong., 2nd sess., August 3, 1964 (Washington, DC: GPO, 1964), 16–39 (truck driver menace), 49 (AMA), 51 ("horrible crimes"), 73–75 (PMA). The next year, Representative Oren Harris, chair of the committee considering the 1965 bill, sarcastically referred back to the "psychotoxic" bill, saying, "Well, under the bill that passed the other body last year, I was advised that even aspirin could be brought under it." See U.S. House, Committee on Interstate and Foreign Commerce, *Drug Abuse Control Amendments of 1965*, 89th Cong., 1st sess., January 27, 28, February 2, 9, 10, 1965 (Washington, DC: GPO, 1965), 92.

44. Sevareid's editorial reported in *F-D-C Reports*, September 14, 1964, 7–13; *CBS Evening News with Walter Cronkite*, September 2–3, 1964, cited in *Drug Abuse Control Amendments of 1965*, 283–87.

45. Public Law 89–74, 79 *Stat.* 227 Sec. 3(a) Section 201 (v) (3). For Isbell's suggestions, see his letters reprinted in U.S. Senate, *Interagency Coordination in Drug Research and Regulation*, 1431–36.

46. HR 130, 89th Cong., 1st sess., 7 (emphasis added).

47. See U.S. Senate, *Interagency Coordination in Drug Research and Regulation*, 92 ("CHAIRMAN: The point is, we have to consider, among other things, the competitive situation—we believe we have the finest system of competitive free enterprise of any nation or any society of people. What is going to be the result to the competitors of the barbiturates if the competitors produce similar results and they are not included?"), 101 ("Representative J. Arthur Younger: Well, you gave as one of the reasons why you did not want to enumerate [each drug] because the manufacturers could not agree with your definition of the drug"), 287–300 (testimony of Dr. V. D. Mattia, executive vice president of Hoffmann-LaRoche). See also *F-D-C Reports*, December 21, 1964, 31–33; February 8, 1965, 16–19; February 15, 1965, 12–15.

48. In its report accompanying the legislation, the House insisted that the secretary of HEW "very soon after the effective date of the legislation, will proceed with the classification as depressant or stimulant of those drugs which are already causing serious problems, primarily certain tranquilizers." HR 130, 89th Cong., 1st sess., 5. The reassurances of Wilbur Cohen, under secretary of the Department of Health, Education, and Welfare, to Senator Ralph Yarborough, June 17, 1965, that his department would act quickly to prevent "unwarranted competitive disadvantage to a firm manufacturing or distributing barbiturates which may be no more habit forming than drugs that are not brought specifically under the control of the act by name—certain tranquilizers, for instance" appeared in the Senate's report, SR 337, 15.

49. Public Law 89–74. For a political/legislative history of these developments, see David Musto and Pamela Korsmeyer, *The Quest for Drug Control: Politics and Federal Policy in a Period of Increasing Substance Abuse, 1963–1981* (New Haven: Yale University Press, 2002), 1–37.

50. 31 *Fed Register* 565, January 18, 1966, and 31 *Fed Register* 4679, March 19, 1966.

51. 31 *Fed Register* 7174, May 17, 1966. A Mr. John Welke of 1130 Haight Street in San Francisco also filed an objection to the listing of LSD and requested a public hearing (he was denied). For the portion of the statute that defines drug abuse, see Section 166.2(e) as written in 31 *Fed Register* 1072.

52. See Hannes Petursson and Malcolm Lader, *Dependence on Tranquilizers* (Oxford: Oxford University Press, 1984), 93, and John Marks, *The Benzodiazepines: Use, Overuse, Misuse, Abuse* (Lancaster, England: MTP, 1985), 34–35.

53. Miltown ads, *JAMA*, October 15, 1955, 38–39, and November 3, 1956, 27.

54. Carter Products clinical bulletin, January 28, 1957, reprinted in U.S. Senate, *Interagency Coordination in Drug Research and Regulation*, 1401.

55. 1958 versions of Miltown advertising brochure, reprinted ibid., 1402.

56. Thomas Haizlip and John Ewing, "Meprobamate Habituation: A Controlled Clinical Study," *New England Journal of Medicine*, June 12, 1958, 1181, 1186. See also Leonard Swanson and Tsuyoshi Okada, "Death after Withdrawal of Meprobamate," *JAMA*, June 8, 1963, 781–82.

57. "Meprobamate," *Medical Letter on Drugs and Therapeutics* 2 (October 28, 1960): 87–88.

58. Lemere, "Drug Habituation," 1431.

59. Walter Osinski, "Withdrawal of Meprobamate," *JAMA*, February 9, 1957, 489.

60. Paul Hoch, "Choice of Sedatives and Tranquilizers," in *Drugs of Choice, 1958–1959* (St. Louis: C. V. Mosby, 1959), 286.

61. Austin Stough, "Possible Habituating Properties of Meprobamate," *JAMA*, February 22, 1958, 882–88.

62. Leo Hollister, Francis Motzenbecker, and Roger Degan, "Withdrawal Reactions from Chlordiazepoxide ('Librium')," *Psychopharmacologia*, 1961, 63–68; Hollister et al., "Diazepam in Newly Admitted Schizophrenics," *Diseases of the Nervous System*, December 1963, 746–50. See Andrea Tone, "Tranquilizers on Trial," in *Medicating Modern America: Prescription Drugs in History*, ed. Andrea Tone and Elizabeth Siegel Watkins (New York: New York University Press, 2007), 168–69.

63. Letter from Alfred F. Zobel to John H. Wood, August 24, 1979, reprinted and discussed in U.S. Senate, Committee on Labor and Human Resources, Subcommittee on Health and Scientific Research, *Use and Misuse of Benzodiazepines*, 96th Cong., 1st sess., September 10, 1979 (Washington, DC: GPO, 1980), 164–72.

64. Barry Maletzky and James Klotter, "Addiction to Diazepam," *International Journal of the Addictions* 11 (1976): 95–115.

65. S. Foster Moore Jr., "Therapy of Psychosomatic Symptoms in Gynecology: An Evaluation of Chlordiazepoxide," *Current Therapeutic Research*, May 1962, 255.

66. T. A. Lamphier, "Drug Addiction," *JAMA*, January 5, 1957, 68.

67. Marks, *Benzodiazepines*, reviewed nearly two hundred case studies of benzodiazepine addiction and found that nearly all of them related to patients with a history of abuse.

68. Stough, "Possible Habituating Properties of Meprobamate," 882, 887.

69. Rotskoff, *Love on the Rocks*; McClellan, "Lady Tipplers"; Nyswander, *Drug Addict as a Patient*.

70. J. A. Ewing and R. E. Fullilove, "Addiction to Meprobamate," *New England Journal of Medicine*, July 11, 1957, 77.

71. Joseph Borrus, "Meprobamate in Psychiatric Disorders," *Medical Clinics of North America*, March 1957, 334–35.

72. Abraham Wikler and William E. Bakewell Jr., "Incidence [of Valium Addiction] in a University Hospital Psychiatric Ward," *JAMA*, May 23, 1966, 712.

73. D. W. Swanson, R. L. Weddige, and R. M. Morse, "Abuse of Prescription Drugs," *Mayo Clinic Proceedings*, May 1973, 365–66.

74. AMA Committee on Alcoholism and Addiction and Council on Mental Health, "Dependence on Barbiturates and Other Sedative Drugs," *JAMA*, August 23, 1965, 674. See also Harris Isbell and H. F. Fraser, "Addiction to Analgesics and Barbiturates," *Pharmacological Reviews* 2 (1950): 355–97, and Acker, *Creating the American Junkie*, 125–55. Not only physicians believed this. See Alcoholics Anonymous, *Tranquilizers, Sedatives, and the Alcoholic* (AA World Service, 1959), which argues that "the same personality pattern that predisposes to addiction, to alcohol and to barbiturates will produce dependence on tranquilizing drugs."

75. *PDR* (Rutherford, NJ: Medical Economics, 1959), 861, 847; *PDR* (1962), 785. A range of other drug reference texts not produced by pharmaceutical companies echoed this reasoning. See, e.g., John Krantz Jr. and C. Jelleff Carr, *The Pharmacologic Principles of Medical Practice* (Baltimore: Williams and Wilkins, 1958), 710; Frederic Flach and Peter Regan, *Chemotherapy in Emotional Disorders: The Psychotherapeutic Use of Somatic Treatments* (New York: McGraw-Hill, 1960), 131; Dale Friend, "Choice of Sedatives and Tranquilizers in General Practice," *in Drugs of Choice, 1962–1963* (St. Louis: C. V. Mosby, 1963), 257.

76. Carl Essig, "Addiction to Nonbarbiturate Sedative and Tranquilizing Drugs," *Clinical Pharmacology and Therapeutics*, May–June 1964, 334–43.

77. *F-D-C Reports*, July 11, 1966, 10–14.

78. For the meprobamate decision, see 32 *Fed Register* 17473, December 6, 1967. For the benzodiazepine decision, see 34 *Fed Register* 7968, May 21, 1969.

79. Essig quoted in *F-D-C Reports*, July 11, 1966, 13; Berger quoted in *F-D-C Reports*, August 1, 1966, 17.

80. 32 *Fed Register* 17474, December 6, 1967.

81. 34 *Fed Register* 7970, May 21, 1969. For the report submitted to the Hearing Examiner, see Exhibit 32, "Memorandum on Librium and Valium Submitted by the Bureau of Narcotics and Dangerous Drugs," and Exhibit 47, "Report on Abuse of Librium and Valium Submitted to the Subcommittee to Investigate Juvenile Delinquency by U.S. Department of Justice, Bureau of Narcotics and Dangerous Drugs," in U.S. Senate, Committee on the Judiciary, Subcommittee to Investigate Juvenile Delinquency, *Narcotics Legislation*, 91st Cong., 1st sess., September 29, 1969 (Washington, DC: GPO, 1969), 649–50, 1167–82.

82. 34 *Fed Register* 2557–58, February 6, 1971.

83. 32 *Fed Register* 17475, December 6, 1967.

84. See, e.g., 32 *Fed Register* 17477, December 6, 1967.

85. See, e.g., 34 *Fed Register* 7971, May 21, 1969.

86. *Carter-Wallace, Inc. v. Gardner,* 417 *F.2d* 1086 (1969); *Hoffmann-La Roche, Inc. v Kleindienst,* 478 *F.2d* (1973).

87. Musto, *American Disease,* 239–40.

88. HR 13742, printed in U.S. House, Committee on Ways and Means, *Legislation to Regulate Controlled Dangerous Substances and Amend Narcotics and Drug Laws,* 91st Cong., 2nd sess. (Washington, DC: GPO, 1970), 23–40 (hereafter cited as 1970 DACA Hearings).

89. Ibid., 280. Amphetamines moved to schedule II in 1971.

90. Ibid., 313, 319. Such reasoning had a long pedigree. In the Kefauver hearings a decade earlier, a physician testified openly that he was maintaining a former barbiturate addict on Miltown. She carried a card, he explained, in case of medical emergencies that might interfere with her regular self-dosing, so that physicians would know to give her the drug and so prevent seizures. U.S. Senate, Committee on the Judiciary, Subcommittee on Antitrust and Monopoly, *Administered Prices in the Drug Industry* (Washington, DC: GPO, 1959–61), 9052.

91. U.S. Senate, Select Committee on Small Business, Subcommittee on Monopoly, *Competitive Problems in the Drug Industry,* 91st Cong., 1st sess., July 16, 29, 30, October 27, 1969 (Washington, DC: GPO, 1969), 5313.

92. 1970 DACA Hearings, 538–40. Ayd stuck to his guns. As late as 1980 he challenged fellow physicians, "Has anyone here ever seen the kind of 'skid row' deterioration from chronic benzodiazepine overuse that develops with alcohol? I have not seen it; and I have looked for it." He claimed to have seen no "major social problems created by it, and I certainly don't see people stealing to obtain it as they do in order to get heroin. I believe such behavior is not found among benzodiazepine users." Frank Ayd, "Social Issues: Misuse and Abuse," *Psychosomatics,* 1980, 22–24.

93. See reports from U.S. House, Committee on Interstate and Foreign Commerce, Subcommittee on Public Health and Environment, *Drug Abuse Prevention Act Oversight,* 92nd Cong., 2d sess., February 1, 2, 4, May 15, 1972 (Washington, DC: GPO, 1972), 167–68 (from *Fed Reg* Doc 71-1664 Filed 2-5-1971 by Ingersoll, BNDD). See, e.g., "Boom in Illegal Pills," *Newsweek,* August 14, 1978, 22.

94. 1970 DACA Hearings, 122.

95. Statement of Dr. V. D. Mattia, president and chief executive officer, Hoffmann-LaRoche, Inc., ibid., 497–99. Earlier, Dr. John Burns, vice president of research at Roche, had suggested that Congress establish different schedules

for narcotics and for "psychotropics." See his testimony in U.S. Senate, *Narcotics Legislation*, 622–23.

96. *WHO Technical Report Series*, 1970, no. 437.

97. Sgt. John R. O'Connor to Mr. Joseph Dooley, in the statement of Dr. V. D. Mattia, Roche president and CEO, 1970 DACA Hearings, 519–20.

98. For a journalistic exposé of Roche's maneuverings, see John Pekkanen, *The American Connection: Profiteering and Politicking in the "Ethical" Drug Industry* (Chicago: Follett, 1973).

99. For similar reasoning about illegal drugs, especially "crack" cocaine, see Nancy Campbell, "Regulating Maternal Instinct: Governing Mentalities of Late-Twentieth-Century U.S. Illicit Drug Policy," *Signs: Journal of Women in Culture and Society* 24, no. 4 (1999): 895–923, and *Using Women: Gender, Drug Policy, and Social Justice* (New York: Routledge, 2000).

Chapter Four: The Valium Panic

1. "Tranquilizers: Use, Abuse, and Dependency," *FDA Consumer*, October 1978, 21.

2. See "Danger Ahead! Valium—The Pill You Love Can Turn on You," *Vogue*, February 1975, 152–53, and "Valium Abuse: The Yellow Peril," *Newsweek*, September 24, 1979, 66. See also, e.g., Penelope McMillan, "Women and Tranquilizers," *Ladies Home Journal*, November 1976, 164–67+; Maya Pines, "Legal Drugs: The Mood Business," *Cosmopolitan*, December 1976, 204–7+; Robert Brewin, "Businessmen Hooked on Valium," *Dun's Review*, January 1978, 44–46; Susan Edmiston, "The Medicine Everybody Loves," *Family Health/Today's Health*, January 1978, 24–28+; "The Three Most Dangerous Drugs," *Good Housekeeping*, March 1978, 233–34; Morris Chafetz and Patrick Young, "The Complete Book of Women and Pills," *Good Housekeeping*, April 1979, 73–76+; Myron Brenton, "Women, Doctors, and Alcohol," *Redbook*, May 1979, 27+; Ellen Switzer, "Inner Info: Your Emotions," *Vogue*, December 1979, 282–83; Elizabeth Whelan and Margaret Sheridan, "The Prescribed Addiction," *Harper's Bazaar*, January 1980, 100–101+; Susan Jacoby, "The Tranquilizer Habit," *McCall's*, January 1980, 42–46; John Hubbell, "Danger! Prescription-Drug Abuse," *Reader's Digest*, April 1980, 100–104; Robert Anson, "Can You Be Just a Little Hooked? The Legal Killers," *Mademoiselle*, October 1980, 224–25+.

3. Barbara Gordon, *I'm Dancing As Fast As I Can* (New York: Harper and Row, 1979). A film version appeared in 1982. Quotations from Patricia Burstein and Barbara Gordon, "TV Producer Barbara Gordon 'Danced' to Valium's Tune—And Landed in a Mental Ward," *People*, June 18, 1979, 98.

4. Conway Hunter, statement before U.S. Senate, Committee on Labor and

Human Resources, Subcommittee on Health and Scientific Research, *Examination on the Use and Misuse of Valium, Librium, and Other Minor Tranquilizers*, 96th Cong., 1st sess., September 10, 1979 (Washington, DC: GPO, 1979), 52. See also U.S. House, Select Committee on Narcotics Abuse and Control, *Abuse of Dangerous Licit and Illicit Drugs: Psychotropics, Phencyclidine (PCP), and Talwin*, 95th Cong., 2nd sess., August 10, 1978 (Washington, DC: GPO 1978); U.S. House, Select Committee on Narcotics Abuse and Control, *Women's Dependency on Prescription Drugs*, 96th Cong., 1st sess., September 13, 1979 (Washington, DC: GPO, 1979), 4–6.

5. The denunciations read like a who's-who of the profession: Leo Hollister at al., "Valium: A Discussion of Current Issues," *Psychosomatics*, January–March 1977, 47; Karl Rickels, "Executive Anxiety and 'Mood' Drugs: Fact vs. Fiction, *Duns Review*, May 1978, 115–16 and "Benzodiazepines: Clinical Use Patterns," *NIDA* 33 (1980): 45; Nathan Kline, "Future of Psychopharmacology," *Minnesota Medicine*, August 1979, 603; Frank Ayd, "Benzodiazepines: Dependence and Withdrawal" [editorial], *JAMA*, September 28, 1979, 1401–2. For a popular overview, see Gilbert Cant, "Valiumania," *New York Times Magazine*, February 1, 1976, 34, 44.

6. Dean Manheimer et al., "Popular Attitudes and Beliefs about Tranquilizers," *American Journal of Psychiatry*, November 1973, 1253; Hugh Parry et al., "National Patterns of Psychotherapeutic Drug Use," *AMA Archives of General Psychiatry*, June 1973, 769; Mitchell Balter and Jerome Levine, "The Nature and Extent of Psychotropic Drug Usage in the United States," *Psychopharmacology Bulletin*, October 1969, 13; Glen Mellinger, Mitchell Balter, and Dean Manheimer, "Patterns of Psychotherapeutic Drug Use among Adults in San Francisco," *AMA Archives of General Psychiatry*, November 1971, 393–94; Balter and Levine, "Character and Extent of Psychotherapeutic Drug Usage in the United States," in *Psychiatry: Proceedings of the V World Congress of Psychiatry, Mexico, D.F., 25 November–4 December, 1971*, ed. Ramón de la Fuente and Maxwell Weisman (New York: American Elsevier, 1973), 87–88.

7. Elizabeth Rasche Gonzáles, "Where Are All the Tranquilizer Junkies?" *JAMA*, May 20, 1983, 2603–4; Jonathan Gabe and Michael Bury, "Tranquilisers and Health Care in Crisis," *Social Science and Medicine* 32, no. 4 (1991): 449–54. See also Gabe, ed., *Understanding Tranquilizer Use: The Role of the Social Sciences* (New York: Routledge, 1991); Mickey Smith, *A Social History of the Minor Tranquilizers: The Quest for Small Comfort in the Age of Anxiety* (New York: Pharmaceutical Products, 1991); Smith, "Lay Periodical Coverage of the Minor Tranquilizers: The First Quarter Century," *Pharmacy in History* 25 (1983): 131–36; Susan Speaker, "From 'Happiness Pills' to 'National Nightmare': Changing Cultural Assessment of Minor Tranquilizers in America, 1955–1980," *Journal of*

the History of Medicine, July 1997, 38–76; Speaker, "Too Many Pills: Patients, Physicians, and the Myth of Overmedication in America, 1955–1980" (Ph.D. diss., University of Pennsylvania, 1992). Historians David Healy and Edward Shorter, writing more broadly about psychiatry and psychopharmacology, do not dismiss the Valium affair in this way. See, e.g., Healy, *The Antidepressant Era* (Cambridge: Harvard University Press, 1997), and Shorter, *A History of Psychiatry: From the Era of the Asylum to the Age of Prozac* (New York: Wiley, 1998).

8. Janice Clinthorne et al., "Changes in Popular Attitudes and Beliefs about Tranquilizers, 1970–1979," *AMA Archives of General Psychiatry* 43 (June 1986): 527–32.

9. David Healy, *The Creation of Psychopharmacology* (Cambridge: Harvard University Press, 2002), 170; Laurence J. Kirmayer, "The Sound of One Hand Clapping: Listening to Prozac in Japan," in *Prozac as a Way of Life*, ed. Carl Elliott and Tod Chambers (Chapel Hill: University of North Carolina Press), 164–93.

10. David Musto, *The American Disease: Origins of Narcotic Control*, 3rd ed. (New York: Oxford University Press, 1999); Nancy D. Campbell, *Using Women: Gender, Drug Policy, and Social Justice* (New York: Routledge, 2000); Caroline Jean Acker, *Creating the American Junkie: Addiction Research in the Classic Era of Narcotic Control* (Baltimore: Johns Hopkins University Press, 2002); Curtis Marez, *Drug Wars: The Political Economy of Narcotics* (Minneapolis: University of Minnesota Press, 2004).

11. See, e.g., John Ratcliff, "The Truth about Sleeping Pills," *Woman's Home Companion*, April 1946, 31, 69; Vera Connolly, "Lethal Lullaby," *Collier's*, October 19, 1946, 86, 95–97; "Sleep-Pill Problem," *Business Week*, March 24, 1945, 88–91. When Minnesota passed a similar law in 1939, incidents of students using barbiturates to get "thrills" percolated through local media to *Newsweek*. See "The 'Lullaby Pill' Peril: Barbiturates, Taken for Binges as Well as Sleep, Attacked," *Newsweek*, March 13, 1939, 36. New York's proposed law generated a *Time* article in 1945. See "Bolts and Jolts," *Time*, October 1, 1945, 94–95. Many of these reports were not sensationalized, and some even defended the barbiturates. See, e.g., Maxine Davis, "Sleeping Pills," *Good Housekeeping*, June 1947, 26–27, 240–43.

12. For articles quoting Isbell, see "Behind the Goofball," *Newsweek*, May 29, 1950, 47; William Engle, "Sleeping Pills: Doorway to Doom," *Coronet*, February 1951, 25–28; Mort Weisinger, "Sleeping Pills Are Worse Than Dope!" *Coronet*, January 1955, 31–35. For other examples from local and regional newspapers, see U.S. House, Committee on Ways and Means, *Traffic in, and Control of, Narcotics, Barbiturates, and Amphetamines*, 84th Cong., 1st sess., October 13, 1955 (Washington, DC: GPO, 1956), 243–57. Defending the barbiturates from

such attacks was Richard Williams, "To Sleep: Perchance . . . ," *Life*, October 13, 1952, 105–6. For an example of a popular report in the wake of congressional hearings, see Ralph Bass, "The Menace of the Sleeping Pill Habit," *Coronet*, March 1958, 100–103. Interestingly, coverage of amphetamines during this early period was far more benign and amused. See, e.g., "Benzedrine: Mental Scores Improve after a Dose of Pills," *Newsweek*, January 2, 1937, 18; "Reducing Made Easy," *Time*, August 23, 1943, 66; "Energy in Pills," *Business Week*, January 15, 1944, 40–42; "Benzedrine Alerts," *Time*, February 21, 1944, 75; "Bolts and Jolts," *Time*, October 1, 1945, 94–95; "FDA Moves to Stamp out Sale of Drugs to Truck Drivers," *Business Week*, October 20, 1945, 42; "Benzedrine for Barbiturates," *Time*, July 1, 1945, 67; "Benzedrine and Dieting," *Newsweek*, September 8, 1947, 48–49; "Jag from Inhalers," *Newsweek*, December 15, 1947, 54; Hannah Lees, "Farewell to Benzedrine Benders," *Collier's*, August 13, 1949, 32, 66; "Benny is My Co-Pilot," *Time*, June 11, 1956, 50.

13. See, e.g., Tom Mahoney, *The Merchants of Life* (New York: Harper, 1959).

14. See chap. 1, n. 16.

15. "Big Pill Bill to Swallow: The Wonder-Drug Makers Get Handsome Profits from Their Captive Consumers," *Life*, February 15, 1960, 97–103. Abbott Laboratories responded by placing a full-page "paid editorial" rebutting the charges. "A Paid Editorial: Report to the Nation's Doctors on 'A Big Pill Bill to Swallow,'" *Life*, March 23, 1960, 121.

16. "Drug Industry: Filling Prescriptions under Fire," *Business Week*, December 10, 1960, 140–54. An extensive collection of media coverage of Kefauver's investigation can be found in "Senate Anti-Trust and Monopoly, Drugs," accession 71a 5170, record group 46, box 6, National Archives Building, Washington DC, ranging from the *Washington Post* to the *Machinist* to Walter Goodman, "Wonder Drugs: How Much Can You Believe?" *Redbook*, February 1960, 31–33+. See also Richard Harris, *The Real Voice* (New York: Macmillan, 1964); Milton Moskowitz, "Wonder Profits in Wonder Drugs," *Nation*, April 27, 1957, 357–60; Harold Jacobson, "The Truth about Prescription Prices," *American Mercury*, July 1957, 19–24; "The High Cost of Rx Drugs," *Consumer Reports*, November 1958, 597–99; "Drugs—The Price You Pay," *Newsweek*, December 7, 1959, 87–89; "Where Congress' Drug Industry Probe Leads," *Business Week*, December 19, 1959, 30–31; David Cowen, "Ethical Drugs and Medical Ethics," *Nation*, December 26, 1959, 479–82; "Investigations: Pills," *Newsweek*, February 1, 1960, 66; "Tranquilizer Makers Put on Spot," *Business Week*, February 6, 1960, 32; "The Truth about Drug Prices," *U.S. News and World Report*, March 21, 1960, 114–19; "Too Many Drugs?" *Time*, April 25, 1960, 78–79.

17. Morton Mintz, *The Therapeutic Nightmare* (Boston: Houghton Mifflin, 1965), reprinted as *By Prescription Only* (Boston: Houghton Mifflin, 1967);

John G. Fuller, *200,000,000 Guinea Pigs: New Dangers in Everyday Foods, Drugs, and Cosmetics* (New York: G. P. Putnam's Sons, 1972), 98–100; John Pekkanen, *The American Connection: Profiteering and Politicking in the "Ethical" Drug Industry* (Chicago: Follett, 1973), 80–81; Milton Silverman and Philip R. Lee, *Pills, Profits, and Politics* (Berkeley: University of California Press, 1974), 58–59; Gerry Stimson, "The Message of Psychotropic Drug Ads," *Journal of Communications,* Summer 1975, 153–60; reports on the National Council of Churches' 1972 Hearings on Drug Advertising, published in *Journal of Drug Issues,* Summer 1974, v–vi, 203–312, and Winter 1976, v–vi, 1–108. These also echoed in medical literature. See, e.g., Richard Feinbloom, "To Open Debate on Tranquilizers," *New England Journal of Medicine,* April 8, 1971, 791; Bruce Ditzion, "Psychotropic Drug Advertisements," *Annals of Internal Medicine* 75 (September 1971): 471; Robert Seidenberg, "Drug Advertising and Perception of Mental Illness," *Mental Hygiene* 55 (January 1971): 21–31, and "Advertising and Abuse of Drugs," *New England Journal of Medicine,* April 8, 1971, 789–90; J. Maurice Rogers, "Drug Abuse—Just What the Doctor Ordered," *Psychology Today,* September 1971, 16–24; Benjamin Gordon, "The Advertising of Psychoactive Drugs," *Journal of Drug Issues,* Winter 1972, 38–41.

18. See, e.g., Harold Schmeck, "Nader Group Sees Pressure on FDA: Asks Congressional Inquiry on Industry's Influence," *New York Times,* April 3, 1972, 30; "The Doctor behind Ralph Nader: What Internist Sidney Wolfe is Doing for—or to—Medicine," *Medical World News,* June 7, 1974, 36–43; "Medical Accountability Sought by Advocate," *U.S. Medicine,* July 15, 1975; Eve Bargmann, Sidney Wolfe, Joan Levin, and the Public Citizen Health Research Group, *Stopping Valium* (Washington, DC: Public Citizen's Health Research Group, 1982).

19. Ritalin ad, *JAMA,* February 8, 1971. See also January 11, 1971; March 8, 1971 ("brownouts"); April 5, 1971 ("peace now!"). For others among many examples, see Roerig's Atarax ad, *JAMA,* May 4, 1957, 79 (nightmares, bedwetting, temper tantrums); Librium ad, *JAMA,* January 7, 1966, back cover ("apprehension" caused by reading medical articles in lay publications).

20. Valium ad, *JAMA,* October 4, 1965, 85; Triavil ad, *JAMA,* January 6, 1975, front cover; Aventyl ad and critique, U.S. Senate, Committee on Small Business, Subcommittee on Monopoly, *Competitive Problems in the Drug Industry, Part 13,* 91st Cong., 1st sess., July 16, 29, 30, October 27, 1969 (Washington, DC: GPO, 1969), 5344–51; Serentil ad, *JAMA,* December 7, 1970. The FDA later required Sandoz Pharmaceuticals to publish a retraction ad, making it clear that Serentil was a powerful antipsychotic not to be used for everyday problems. See Serentil advertisement/retraction, *JAMA,* April 5, 1971.

21. Robert Seidenberg, in U.S. Senate, Select Committee on Small Business, Subcommittee on Monopoly, *Advertising of Proprietary Medicines, Part 2: Mood*

Drugs (Sedatives, Tranquilizers, and Stimulants), 92nd Cong., 1st sess., July 21, 22, 23, September 22, 1971 (Washington, DC: GPO, 1971), 542.

22. See, e.g., Hugh Parry, "Use of Psychotropic Drugs by U.S. Adults," *Public Health Reports* 83 (October 1968): 799–810, and Parry et al., "National Patterns of Psychotherapeutic Drug Use."

23. Reginald Bowes, "The Industry as Pusher," *Journal of Drug Issues*, Summer 1974, 238. See also Pam Gorring, "Multinationals or Mafia: Who Really Pushes Drugs?" in *Two Faces of Deviance: Crimes of the Powerless and Powerful*, ed. P. R. Wilson and J. Braithwaite (Brisbane, Australia: University of Queensland Press, 1978), 82.

24. Paul Starr, *The Social Transformation of American Medicine: The Rise of a Sovereign Profession and the Making of a Vast Industry* (New York: Basic Books, 1982), 379–420, esp. 389.

25. Sara Evans, *Tidal Waves: How Women Changed America at Century's End* (New York: Free Press, 2004).

26. Sandra Morgen, *Into Our Own Hands: The Women's Health Movement in the United States, 1969–1990* (New Brunswick, NJ: Rutgers University Press, 2002), 1–40. See also Sheryl Burt Ruzek, *The Women's Health Movement: Feminist Alternatives to Medical Control* (New York: Praeger, 1978). Classic texts include Boston Women's Health Book Collective, *Our Bodies, Ourselves* (Boston: New England Free Press, 1971); Barbara Ehrenreich and Deirdre English, *Complaints and Disorders: The Sexual Politics of Sickness* (Old Westbury, NY: Feminist Press, 1973); Germaine Greer, *The Female Eunuch* (London: MacGibbon and Kee, 1971); Jean Baker Miller, ed., *Psychoanalysis and Women* (New York: Penguin Books, 1973); Phyllis Chesler, *Women and Madness* (Garden City, NY: Doubleday, 1972).

27. Morgen, *Into Our Own Hands*, 1–40.

28. Anthony Lukas, "The Drug Scene: Dependence Grows," *New York Times*, January 8, 1968, 22.

29. Joel Fort, *The Pleasure Seekers: The Drug Crisis, Youth, and Society* (Indianapolis: Bobbs-Merrill, 1969), 195–96; Fort, "Why People Use Drugs and What Should Be Done about It," in *Progress in Drug Abuse: Proceedings of the Third Annual Western Institute of Drug Problems Summer School*, ed. Paul H. Blachly (Springfield, IL: Charles C Thomas, 1970), 10–26. See also Fort and Christopher Cory, *American Drugstore: A (Alcohol) to V (Valium)* (Boston: Educational Associates, 1975); "Addiction: Fact and Fiction," *Saturday Review*, August 18, 1962, 30–31; "AMA Lies about Pot," *Ramparts*, August 24, 1968, 12+; "Drug Use and the Law," *Current*, December 1969, 4–13.

30. William White, *Slaying the Dragon: The History of Addiction Treat-*

ment and Recovery in America (Bloomington, IL: Chestnut Health Systems/ Lighthouse Institute, 1998), 263–78.

31. E.g., Reginald Smart and Dianne Fejer, "Drug Abuse among Adolescents and Their Parents: Closing the Generation Gap in Mood Modification," *Journal of Abnormal Psychology* 79, no. 2 (1972): 153–60. In the late 1960s the Food and Drug Administration pushed pharmaceutical companies to rein in their advertising, especially for over-the-counter sleep and energy aids. As one congressman told the *New York Times* in 1970, "These television ads were creating an acceptance of pills among the very young viewers, and this lowered their resistance to taking illegal drugs." "Drug Spots on TV to Be Toned Down," *New York Times*, September 18, 1970.

32. U.S. Senate, *Advertising of Proprietary Medicines*, 926–27.

33. Edward Brecher, *Licit and Illicit Drugs: The Consumers Union Report on Narcotics, Stimulants, Depressants, Inhalants, Hallucinogens, and Marijuana— Including Caffeine, Nicotine, and Alcohol* (Mount Vernon, NY: Consumers Union, 1972).

34. Roland Berg, "Drugs: The Mounting Menace of Abuse," *Look*, August 8, 1967, 11–28.

35. The expanding literature on white suburban racial and economic anxieties includes Matthew Lassiter, *The Silent Majority: Suburban Politics in the Sunbelt South* (Princeton: Princeton University Press, 2005); Eric Avila, *Popular Culture in the Age of White Flight: Fear and Fantasy in Suburban Los Angeles* (Berkeley: University of California Press, 2006); Thomas Sugrue, *Origins of the Urban Crisis: Race and Inequality in Postwar Detroit* (Princeton: Princeton University Press, 1995); Lisa McGirr, *Suburban Warriors: The Origins of the New American Right* (Princeton: Princeton University Press, 2002).

36. "Drug Culture: Take a Look at Your Office," *Business Week*, August 15, 1970, 83.

37. Martin Arnold, "The Drug Scene: A Growing Number of America's Elite Are Quietly 'Turning On,'" *New York Times*, January 10, 1968, 26.

38. Carl Chambers and Dodi Schultz, "Women and Drugs," *Ladies Home Journal*, November 1971, 191.

39. Sam Blum, "Pills That Make You Feel Better," *Redbook*, August 1968, 125.

40. Arnold, "Drug Scene."

41. Earl Gottschalk, "The Perils of Pill-Popping with Mood Drugs," *Science Digest*, July 1969, 13–17, reprinted from the *Wall Street Journal*.

42. *Ladies Home Journal*, December 1971, 66–68.

43. Blum, "Pills That Make You Feel Better."

44. Berg, "Drugs: The Mounting Menace of Abuse."

45. Carl Chambers and Dodi, Schultz, "Housewives and the Drug Habit," *Ladies Home Journal,* December 1971, 138.

46. Bruce Jackson, "White-Collar Pill Party," *The Atlantic,* August 1966, 35–40, reprinted in *Observations of Deviance,* ed. Jack Douglas (New York: Random House, 1970), 256–59.

47. Lukas, "Drug Scene: Dependence Grows."

48. Joan Didion, *Play It As It Lays: A Novel* (New York: Farrar, Straus, and Giroux, 1970); Jacquelyn Susann, *Valley of the Dolls* (New York: Bantam Books, 1966).

49. Barry Blackwell, "Psychotropic Drugs in Use Today: The Role of Diazepam in Medical Practice," *JAMA,* September 24, 1973, 1640.

50. Refill numbers were tracked by the trade journal *American Druggist*'s fortnightly prescription surveys.

51. U.S. Department of Commerce, *Drug Utilization in the U.S.—1985: Seventh Annual Review* (Springfield, VA: National Technical Information Service, 1986).

52. Parry et al., "National Patterns of Psychotherapeutic Drug Use."

53. See, e.g., ABC and NBC evening television news broadcasts for August 15, 1973 (available from the Vanderbilt Television News Archive (http://tvnews.van derbilt.edu/, accessed May 20, 2007); "U.S. Ready to Propose Controls on Use of Librium and Valium," *New York Times,* January 31, 1975, 57; "The Growing Debate over Safety of Drugs," *U.S. News and World Report,* June 16, 1975, 61–62.

54. See "Federal Agency Lists Most Widely Abused Drugs," *JAMA,* August 2, 1976, 432. Also Philip Person, "The Drug Abuse Warning Network: A Statistical Perspective," *Public Health Reports,* September–October 1976, 395–402. The news was reported, e.g., on *NBC Evening News* for Thursday, July 8, 1976 (from the Vanderbilt Television News Archive, http://tvnews.vanderbilt.edu/, accessed May 20, 2007).

55. "Danger Ahead! Valium—The Pill You Love Can Turn on You," *Vogue,* February 1975, 152–53. See also Nyswander in John Lombardi, "Valium: The Popcorn of the 1970s," *Oui,* September 1976, 96, and McMillan, "Women and Tranquilizers." An article in *Working Woman* quoted an addiction specialist: "I have a woman in my detox unit right now. She's hallucinating; she's psychotic; she's repeating simple phrases over and over again. We've now determined that she has definite, organic brain damage." Sheila Eby, "Stress-Fighting Drugs: Do They Affect Job Competence?" *Working Woman,* April 1980, 10.

56. Richard Hughes and Robert Brewin, *The Tranquilizing of America: Pill-Popping and the American Way of Life* (New York: Harcourt Brace Jovanovich, 1979); Eve Bargmann, Sidney Wolfe, Joan Levin, and the Public Citizens Health

Research Group, *Stopping Valium* (Washington, DC: Public Citizens Health Research Group, 1982).

57. See, e.g., Emily Martin, *The Woman in the Body* (Boston: Beacon Press, 1987); Inge Broverman et al., "Sex Role Stereotypes and Clinical Judgments of Mental Health," *Journal of Consulting and Clinical Psychology*, February 1970, 1–7; Jean and John Lennane, "Alleged Psychogenic Disorders in Women," *New England Journal of Medicine*, February 8, 1973, 288–91; Ruth Cooperstock, "Sex Differences in the Use of Mood-Modifying Drugs: An Explanatory Model," *Journal of Health and Social Behavior*, September 1971, 238–44; Mary Howell, "What Medical Schools Teach about Women," *New England Journal of Medicine*, August 8, 1974, 304–7.

58. Elissa Henderson Mosher, "Portrayal of Women in Drug Advertising: A Medical Betrayal," *Journal of Drug Issues* 6, no. 1 (1976): 72–78 at 74. Mady Schutzman's gender analysis of advertisements and hysteria, *The Real Thing: Performance, Hysteria, and Advertising* (Hanover, NH: University Press of New England, 1999), found similar dynamics in advertisements from the nineteenth century to the present. Such ads, she argues, sought to coopt women in the unrealizable project of "fixing" what advertisements themselves defined as the unacceptable essence of womanhood. This circular dialectic parallels the broader project of drug advertisers, who sought to define both male and female consciousness as essentially ill in modern society, and then to sell products to "fix" that illness (see chap. 2).

59. For a recent explication of the medicalization argument, see Margot Lyon, "C. Wright Mills Meets Prozac: The Relevance of 'Social Emotion' to the Sociology of Health and Illness," in *Health and the Sociology of Emotions*, ed. Veronica James and Jonathan Gabe (Oxford: Blackwell, 1996), 55–78. For earlier theoretical discussions of medicalization, see Howard Waitzkin and Barbara Waterman, *The Exploitation of Illness in Capitalist Society* (Indianapolis: Bobbs-Merrill, 1981), and Irving Kenneth Zola, "Medicine as an Instrument of Social Control," in *The Sociology of Health and Illness: Critical Perspectives*, ed. Peter Conrad and Rochelle Kern (New York: St. Martin's, 1980).

60. Robert Seidenberg, testimony before U.S. Senate, *Advertising of Proprietary Medicines*, 551. For similar criticism see also, e.g., F. Suffet and R. Brotman, "Female Drug Use: Some Observations," *International Journal of Addictions*, 1976, 31; Pauline Bart, "Depression in Middle-Aged Women," and Phyllis Chesler, "Patient and Patriarch: Women in the Psychotherapeutic Relationship," in *Woman in Sexist Society: Studies in Power and Powerlessness*, ed. Vivian Gornick and Barbara Moran (New York: Basic Books, 1971), 163–86 and 362–89; Walter Gove and Jeannette Tudor, "Adult Sex Roles and Mental Illness," *American*

Journal of Sociology, 1973, 812–35; Daniel Zwerdling, "Pills, Profits, Peoples' Problems," *Progressive*, October 1973, 44–47; Christine McRee, Billie Corder, and Thomas Haizlip, "Psychiatrists' Responses to Sexual Bias in Pharmaceutical Advertising," *American Journal of Psychiatry*, November 1974, 1273–75; Jane Prather and Linda Fidell, "Sex Differences in the Content and Style of Medical Advertisements," *Social Science and Medicine* 9 (1975): 23–26; Andrea Mant and Dorothy Broom Darroch, "Media Images and Medical Images," *Social Science and Medicine* 9 (1975): 613–18; Deborah Larned, "The Selling of Valium," *Ms.*, November 1975, 32–33; Gerry Stimson, "Women in a Doctored World," *New Society*, May 1, 1975, 265–67; Mosher, "Portrayal of Women in Drug Advertising"; Mickey C. Smith and Lisa Griffin, "Rationality of Appeals Used in the Promotion of Psychotropic Drugs: A Comparison of Male and Female Models," *Social Science and Medicine* 11 (1977): 409–14; Ellie King, "Sex Bias in Psychotropic Drug Advertisements," *Psychiatry*, May 1980, 129–37; Tona Kiefer, "The 'Neurotic Woman' Syndrome," *Progressive*, December 1980, 23–29. Less directly political analysis came from Canadian sociologist Ruth Cooperstock and some others who argued that women were more attuned to their emotional lives, and were also in more regular contact with physicians, and that prescriptions for drugs were a predictable (and positive) consequence. Men, on the other hand, were conditioned against admitting emotional distress and tended to seek release in other culturally sanctioned venues such as drinking in bars. (Indeed, the research team pointedly questioned whether it was in fact men who used too few of these drugs rather than women who used too many.) See Cooperstock's edited volume *Social Aspects of Medical Use of Psychotropic Drugs* (Toronto: Addiction Research Foundation, 1974) as well as her articles "Psychotropic Drug Use among Women," *Canadian Medical Journal*, October 23, 1976, 760–63; "Sex Differences in Psychotropic Drug Use," *Social Science and Medicine* 12B (1978): 179–86; "A Review of Women's Psychotropic Drug Use," *Canadian Journal of Psychiatry* 24, no. 1 (1979): 29–34. See also Derek Phillips and Bernard Segal, "Sexual Status and Psychiatric Symptoms," *American Sociological Review*, February 1969, 58–72; Allan Horwitz, "The Pathways into Psychiatric Treatment: Some Differences between Men and Women," *Journal of Health and Social Behavior*, 1977, 169–78; Ronald C. Kessler, Roger L. Brown, and Clifford L. Broman, "Sex Differences in Psychiatric Help Seeking," *Journal of Health and Social Behavior*, 1981, 49–64.

61. Suzanne Levine and Harriet Lyons, *The Decade of Women: A Ms. History of the Seventies in Words and Pictures* (New York: Putnam, 1980).

62. Whelan and Sheridan, in "Prescribed Addiction," 100, wrote: "These legal medications . . . are also the chief source of drug abuse among one million women (known as 'hidden addicts') in the U.S. today." Kiefer, " 'Neurotic Woman'

Syndrome"; Muriel Nellis, *The Female Fix* (New York: Penguin, 1980). Marvin Grosswirth, "Valium Danger Signs," *Science Digest,* March 1980, 40–45, even referred to the anti-Valium ranks as "feminist advocates."

63. Betty Ford with Chris Chase, *Betty: A Glad Awakening* (Garden City, NY: Doubleday, 1987), 39–50. See also "Prisoner of Pills," *Newsweek,* April 24, 1978; "Betty's Ordeal," *Time,* April 24, 1978; "Trouble: Betty Ford and Drugs," *People,* April 24, 1978, 49; Byra MacPherson, "Betty Ford: The Untold Story," *McCall's,* July 1978, 18+; "Mrs. Ford's Hospital Stay Linked to Medications," *New York Times,* April 12, 1978, A13; "Mrs. Ford's Illness Described; 'She Hasn't Failed Yet,'" *New York Times,* April 16, 1978, 24; "Abuse of Prescription Drugs: A Serious but Hidden Problem for Women," *New York Times,* April 19, 1978, A12. An autobiography and two biographies published shortly thereafter dealt with her substance abuse. Betty Ford with Chris Chase, *The Times of My Life* (New York: Harper and Row, 1978); Bruce Cassiday, *Betty Ford: Woman of Courage* (New York: Dale, 1978); Sheila Weidenfeld, *First Lady's Lady: With the Fords at the White House* (New York: G. Putnam's Sons, 1979).

64. See, e.g., "The Prisoners of Pills," *Newsweek,* April 24, 1978, 77; quotation from "Betty's Ordeal," *Time,* April 24, 1978, 31; "In Trouble," *People,* April 24, 1978, 49; "In Her Own Words," *People,* May 8, 1978, 102–4+; "Betty Ford: The Untold Story," *McCall's,* July 1978, 18–24, 144.

65. "In Her Own Words," *People,* May 8, 1978, 102.

66. *McCall's,* for example, reported that a Cleveland treatment program received more than one hundred calls the day after Ford's press conference—a tenfold increase from its usual pace. "Betty Ford," *McCall's,* July 1978, 18.

67. Gordon, *Dancing As Fast As I Can,* 16.

68. Ibid.

69. See, e.g., U.S. Senate, Committee on Labor and Human Resources, Subcommittee on Health and Scientific Research, *Examination on the Use and Misuse of Valium, Librium, and Other Minor Tranquilizers,* 11; Hughes and Brewin, *Tranquilizing of America,* 70; "They're Finding a Way Out of the Prescription Drug Trap," *New York Times,* February 3, 1979, 12; Chafetz and Young, "Complete Book of Women and Pills,"; Brenton, "Women, Doctors, and Alcohol."

70. Barbara Gordon, "In Her Own Words: Addicted to Valium," *People,* June 18, 1979, 92, 97–98. The muckraking book *The Tranquilizing of America* introduced what the authors called "an ideal patient—female, white, middle class, college educated, who trusted doctors." She became addicted to Valium despite taking no more than the prescribed dose. Only after bravely confronting her doctor was she able to extract his promise to stop pushing Valium on her. As the authors concluded approvingly, she was "no longer the ideal patient." Hughes and Brewin, *Tranquilizing of America,* 70. See also "They're Finding a Way out of

the Prescription Drug Trap," *New York Times*, February 3, 1979, 12; Chafetz and Young, "Complete Book of Women and Pills"; Brenton, "Women, Doctors, and Alcohol."

71. "In Her Own Words," *People*, May 8, 1978, 106.

72. Jacoby, "Tranquilizer Habit"; "The Three Most Dangerous Drugs," *Good Housekeeping*, March 1978, 233–34; Whelan and Sheridan, "Prescribed Addiction."

73. U.S. House, *Women's Dependency on Prescription Drugs*, 4–7. For another similar program, see information presented about Drug Liberation Program, which had a "Women's Center" devoted to women prescription drug abusers, in U.S. Senate, Committee on Labor and Human Resources, Subcommittee on Health and Scientific Research, *Use and Misuse of Benzodiazepines*, 96th Cong., 1st sess., September 10, 1979 (Washington, DC: GPO, 1980), 505–19.

74. See, e.g., Hannes Petursson and Malcolm Lader, *Dependence on Tranquilizers* (Oxford: Oxford University Press, 1984), 93, and John Marks, *The Benzodiazepines: Use, Overuse, Misuse, Abuse* (Lancaster, England: MTP, 1985), 34–35.

75. Glen Mellinger and Mitchell Balter, "Prevalence and Patterns of Use of Psychotherapeutic Drugs: Results from a 1979 National Survey of American Adults," in *Epidemiological Impact of Psychotropic Drugs*, ed. C. Tognoni, C. Bellantuono, and M. Lader (New York: Elsevier/North-Holland Biomedical Press, 1981), 117–35. See also, e.g., David Greenblatt, Richard Shader, and Jan Koch-Weser, "Psychotropic Drug Use in the Boston Area: A Report from the Boston Collaborative Drug Surveillance Program," *AMA Archives of General Psychiatry*, April 1975, 519, which found that 4.5% of all new patients surveyed had been regularly taking minor tranquilizers for over a year.

76. The decline began, gradually, after Valium appeared on the list of controlled substances in 1975 but grew noticeably more intense at the end of the decade. See, e.g., U.S. Department of Commerce, *Drug Utilization in the U.S.—1985*. Clinthorne et al., "Changes in Popular Attitudes and Beliefs about Tranquilizers, 1970–1979."

77. Valium ads, *JAMA*, front cover, October 6, 1978; November 17, 1978; December 1, 1978; Valium ad, *Medical Times*, March 1981, 17–19.

78. See, e.g., *Medical World News*, May 1982, 20–22: Tranxene "helps avoid effects that encourage drug-seeking behavior" and "helps avoid drug-induced 'buzz' or 'high,'" and so is the perfect drug "if you're concerned about tranquilizer effects that promote unwarranted requests for prescriptions."

79. Peter Kramer, *Listening to Prozac: A Psychiatrist Explores Antidepressant Drugs and the Remaking of the Self* (New York: Viking, 1993).

80. "Tranquilizers: Use, Abuse, and Dependency," *FDA Consumer*, October

1978, 21. See also "The Three Most Dangerous Drugs," *Good Housekeeping*, March 1978, 233–34; Hughes and Brewin, *Tranquilizing of America*, 57–59 ("Surrounded by doctors and given the 'best medical care' in her community, Sharon West, a well-educated white suburban mother of two, became first a prescription junkie and later an alcoholic"). For a rare story about a man, see Brewin, "Businessmen Hooked On Valium," 44.

81. U.S. House, *Women's Dependency on Prescription Drugs*, 4–6.

82. Ibid., 12.

83. George Lipsitz, *The Possessive Investment in Whiteness: How White People Profit from Identity Politics* (Philadelphia: Temple University Press, 1998).

84. See, e.g., Clinthorne et al., "Changes in Popular Attitudes and Beliefs about Tranquilizers, 1970–1979." The authors reported a sharp increase in distrust of the medications and in feelings that they were overprescribed—although the willingness to actually take the drugs to ease suffering remained relatively unaffected.

85. Interestingly, this was also the time when Freud came under the most intense and successful criticism in America. Again, this suggests that the linkages drawn by some historians between the decline of Freud and the rise of psychopharmaceuticals are overdrawn. In the 1970s, *both* were under serious attack.

Chapter Five: Prozac and the Incorporation of the Brain

1. Peter Kramer, *Listening to Prozac: A Psychiatrist Explores Antidepressant Drugs and the Remaking of the Self* (New York: Viking, 1993).

2. Leslie Farber, "Merchandising Depression," *Psychology Today*, April 1979, 63. See also, e.g., Arnold Hutschnecker, "If You're Depressed," *Vogue*, January 15, 1972, 52; "The Most Common Mental Disorder," *Time*, August 7, 1972, 16; Linda Wolfe, "When the Blues Don't Go Away," *McCall's*, December 1972, 132; "Coping with Depression," *Newsweek*, January 8, 1973, 51; Rona Cherry and Laurence Cherry, "The Common Cold of Mental Ailments: Depression," *New York Times Sunday Magazine*, November 25, 1973, x, 117–19, 126–27, 134–35; Gerald Knox, "Blues Really Get You Down," *Better Homes and Gardens*, January 1974, 12–18 (which began, "If the Fifties were the age of anxiety . . . the Seventies are the age of melancholy"); "What You Should Know about Mental Depression," *U.S. News and World Report*, September 9, 1974, 37–40; "Edwin E. 'Buzz' Aldrin: 'I've Been There': A Candid Interview with This Year's National Mental Health Chairman," *Today's Health*, Winter 1974, 4–5; "New Ways to Treat Depression," *Good Housekeeping*, July 1975, 149; Anthony Wolff, "Medicine for Melancholy," *Saturday Review*, February 21, 1976, 34; Theodore Rubin, M.D., "Psychiatrist's Notebook," *Ladies Home Journal*, May 1976, 26; Alice Kosner,

"What to Do When You're *Really* Depressed," *McCall's*, November 1977, 220–21+; Jane Brody, "Mental Depression: The Recurring Nightmare," *New York Times*, January 19, 1977, 56; "The Depression Epidemic," *USA Today Magazine*, April 1979, 15–16; Marilyn Mercer and E. J. Sachar, "The Complete Book of Depression," *Good Housekeeping*, October 1979, 91+.

3. Stanley Jackson, *Melancholia and Depression: From Hippocratic Times to Modern Times* (New Haven: Yale University Press, 1986).

4. See Committee on Nomenclature and Statistics of the APA, *Diagnostic and Statistical Manual of Mental Disorders* (Washington, DC: American Psychiatric Association, 1952), 33–34.

5. The *National Disease and Therapeutic Index* estimated that 7.4 million Americans were diagnosed as having some form of psychosis in 1962, ranking it relatively low on the list of most common illnesses. Of these 7.4 million, evidence suggests that anywhere from 5% to 50% may have received one of the "depressive" labels. *National Disease and Therapeutic Index* (Ambler, PA: Lea and Associates, 1962). See Roy Grinker et al., *The Phenomena of Depressions* (New York: Harper and Row, 1961), for figures on psychotic depression and a complaint that "very few investigations have been made on this syndrome as contrasted with the intensive work carried out on schizophrenic, psychosomatic, and other psychiatric conditions" (xi). For other estimates of depression's prevalence in the 1950s and 1960s, see Charlotte Silverman, *The Epidemiology of Depression* (Baltimore: Johns Hopkins Press, 1968). An overview emphasizing depression's rarity at this time can be found in David Healy et al., "The Burden of Psychiatric Morbidity," *Psychological Medicine* 31 (2001): 779–90, plus quick references in Healy's *The Creation of Psychopharmacology* (Cambridge: Harvard University Press, 2002), 57, 69.

6. Nicolas Rasmussen, "Making the First Anti-Depressant: Amphetamine in American Medicine, 1929–1950," *Journal of the History of Medicine and Allied Sciences* 61(3), 2006, 288–323. *Diagnostic and Statistical Manual of Mental Disorders*, 33–34. This interpretation also appeared, e.g., in Jack Ewalt and Dana Farnsworth, *Textbook of Psychiatry* (New York: Blakiston Division, McGraw-Hill, 1963), 112–13, and Jackson Smith, *Psychiatry: Descriptive and Dynamic* (Baltimore: Williams and Wilkins, 1960), 100–101.

7. *National Disease and Therapeutic Index* (1962).

8. See, e.g., Joseph Tobin and Nolan Lewis, "New Psychotherapeutic Agent, Chlordiazepoxide," *JAMA*, November 5, 1960, 1242; Henry Cromwell, "Controlled Evaluation of Psychotherapeutic Drug in Internal Medicine," *Clinical Medicine*, December 1963, 2239–44; Winston Burdine, "Diazepam in General Psychiatric Practice," *American Journal of Psychiatry*, December 1964, 589–92; *Miltown: The Tranquilizer with Muscle Relaxant Action* (New Brunswick, NJ:

Wallace Laboratories, 1958), 17–19. For utilization patterns, see Karl Rickels, "Pharmacotherapy of Depression," *Psychosomatics,* September–October 1962, 390–98, and Seymour Baron and Seymour Fisher, "Use of Psychotropic Drug Prescriptions in a Prepaid Group Practice Plan," *Public Health Reports,* October 1962, 871–81. Similar results are reported in Sam Shapiro and Seymour Baron, "Prescriptions for Psychotropic Drugs in a Noninstitutional Population," *Public Health Reports,* June 1961, 481–88.

9. Rasmussen argues persuasively based on production figures that amphetamines were widely popular as treatments for mild depression in the 1940s; see "Making the First Anti-depressant." By the 1950s, however, the push to establish amphetamines as antidepressants appears to have faded. Two early prescription drug studies, for example, found that in 1958 Miltown was prescribed more often than amphetamines for depression. The vast majority of amphetamines, meanwhile, were prescribed for obesity and other "nutritional conditions," suggesting limited success in identifying them as antidepressants. See Seymour Baron and Seymour Fisher, "Use of Psychotropic Drug Prescriptions in a Prepaid Group Practice Plan," *Public Health Reports,* October 1962, 874; Sam Shapiro and Seymour Baron, "Prescriptions for Psychotropic Drugs in a Noninstitutional Population," *Public Health Reports,* June 1961, 481–88.

10. Elavil ad, *American Journal of Psychiatry,* April 1964; Marplan ad, *American Journal of Psychiatry,* July 1962; Niamid ad, *American Journal of Psychiatry,* July 1962.

11. A. J. Malerstein, "Depression as a Pivotal Affect," *American Journal of Psychotherapy,* 1968, 22, 202. See also Raymond Friedman, "The Psychology of Depression: An Overview," in *The Psychology of Depression: Contemporary Theory and Research,* ed. Raymond Friedman and Martin Katz (Washington, DC: V. H. Winston and Sons, 1974), 283–84, which includes a bibliography of other pioneering thinkers: "The history of the affect of depression is an intriguing one because for many years the affect was considered a 'second-class citizen,' the step-child of anxiety," the author maintained; "Freud concentrated on anxiety, making it the pivotal affect not only of the psychoneuroses but of all psychic symptomatology. Depression was regarded as merely a defense against anxiety, and this theoretical prejudice remains influential on the contemporary scene."

12. Frank Ayd, *Recognizing the Depressed Patient* (New York: Grune and Stratton, 1961).

13. Nathan Kline, "The Practical Management of Depression" [Lasker Award acceptance speech], *JAMA,* November 23, 1964, 122–30, and "Medicine: Lasker Awards," *New York Times,* November 22, 1964, E6. Similar views from Kline in "Psychopharmaceuticals: Uses and Abuses," *Postgraduate Medicine,* May 1960, 620–29; "The Use of Psychopharmaceuticals in Office Practice," *Medical Clin-*

ics of North America, November 1961, 1677–84; *Depression: Its Diagnosis and Treatment* (New York: S. Karger, 1969).

14. For the founding texts of these schools, see Aaron T. Beck, *Depression: Causes and Treatment* (Philadelphia: University of Pennsylvania Press, 1967), and Myrna Weissman and Eugene Paykel, *The Depressed Woman: A Study of Social Relationships* (Chicago: University of Chicago Press, 1974).

15. See, e.g., Frank McGowan, "The Doctor Talks about Depression," *McCall's*, May 1958, 151, 153; Karl Huber, "The Blues and How to Chase Them," *McCall's*, August 1960, 98; Lawrence Galton, "Depression," *Popular Science*, December 1962, 79–82; Bryant Roisum, "How to Recognize Suicidal Depression," *Ladies Home Journal*, September 1964, 26–27; Janet Graham, "To Beat the Blues," *Reader's Digest*, April 1967, 39–40+; Leonard Cammer, "A Psychiatrist Talks about Depression," *Ladies Home Journal*, February 1969, 60–61; "How to Beat the Blues," *Better Homes and Gardens*, February 1970, 116–17, 122; Huschnecker, "If You're Depressed," 52–54+. Two popular-audience books on depression also appeared: Freud popularizer Lucy Freeman's *Cry For Love: Understanding and Overcoming Human Depression* (Toronto, Ontario: Collier-Macmillan Canada, 1969) and biological psychiatrist Leonard Cammer's *Up from Depression* (New York: Simon and Schuster, 1969).

16. "McGovern's First Crisis: The Eagleton Affair," *Time*, August 7, 1972, 11–16. A partial transcript of his July 25 press conference (at which McGovern indicated full support for him) appears in "Eagleton's Own Story of His Health Problems," *U.S. News and World Report*, August 7, 1972, 16–17. For an overview of the Eagleton affair, see Lawrence Strout, "Politics and Mental Illness: The Campaigns of Thomas Eagleton and Lawton Chiles," *Journal of American Culture* 18, no. 3 (1995): 67–73.

17. Royce Rensberger, "Psychiatrists Explain Medical Facts in Depression Controversy," *New York Times*, July 29, 1972, 11, which also cited figures from the National Institute of Mental Health estimating that four to eight million Americans suffered from medically significant depression. See Dean Schuyler and Martin Katz, *The Depressive Illnesses: A Major Public Health Problem* (Rockville, MD: National Institute of Mental Health, 1973). For other examples, see "McGovern's First Crisis," *Time*, August 7, 1972, 11–16; "Depression and Electroshock," *Newsweek*, August 7, 1972, 20; "Coping with Depression," *Newsweek*, January 8, 1972, 51–54; Wolfe, "When the Blues Don't Go Away"; "Blues Really Get You Down," *Better Homes and Gardens*, January 1974, 12–18; "What You Should Know about Mental Depression," *U.S. News and World Report*, September 9, 1974, 37–40; Kosner, "What to Do When You're *Really* Depressed"; Mercer and Sachar, "Complete Book of Depression."

18. Library copies of Ayd's book still have the Merck sticker on the inside cover. See also David Healy, *Let Them Eat Prozac: The Unhealthy Relationship between the Pharmaceutical Industry and Depression* (New York: New York University Press, 2004), 8. E.g., Pfizer launched its combination drug Sinequan with this proud claim: "First, there were tranquilizers for anxiety . . . Then, there were antidepressants for depression . . . Now, there is NEW Sinequan, the antidepressant that is a tranquilizer . . . Active throughout the spectrum of psychoneurotic anxiety/depression." Sinequan ad, *JAMA*, November 10, 1969. For amphetamine advertising in the 1940s, see Rasmussen, "Making the First Antidepressant."

19. Deprol ad, *JAMA*, November 9, 1963, 74–75. A later ad claimed that "non-psychotic depression and anxious depression—the types of depression most commonly seen and treated in general office practice—respond to 'Deprol' in three out of four patients. Thus, more potent agents can often be reserved for the most difficult cases." Deprol ad, *JAMA*, September 19, 1966. Deprol ads in the *American Journal of Psychiatry* were also pushier, following the same pictorial format but emphasizing the drug's value as an aid to psychotherapy. One ad reads, "RAPPORT! 'My husband? All he does is complain . . . says I'm no fun to live with. I try to snap out of it but I can't seem to. What more can I do?" Deprol ad, *American Journal of Psychiatry*, December 1963.

20. Triavil ad, *JAMA*, September 27, 1965.

21. Triavil ad, *JAMA*, November 3, 1969; Elavil ad, *JAMA*, September 27, 1965, 250–51. See also Triavil ad, *JAMA*, September 27, 1965, 63–67; Aventyl ad, *JAMA*, February 7, 1966.

22. Elavil ad, *JAMA*, December 19, 1966.

23. Triavil ad, *JAMA*, June 21, 1971.

24. Elavil ad, *JAMA*, October 13, 1969. See also, e.g., Triavil ad, *JAMA*, September 11, 1967 ("Probing beneath the surface of anxiety can reveal coexisting depression"); Etrafon ad, *JAMA*, March 11, 1968, 75–82 ("When you recognize, as you often will, the sadness concealed behind the superficial smile or jocularity, the unconscious self-defense behind vague somatic complaints, the anxiety behind hopelessness and fatigue, the depression behind restless activity, the revealing pattern of insomnia, anorexia, loss of interest—you will see that the patient's true problem is very often coexisting depression and anxiety"). Aventyl ads, *JAMA*, February 26, 1968; March 11, 1968; April 8, 1968; May 13, 1968.

25. Aventyl ads, *JAMA*, February 26, 1968; March 11, 1968; April 8, 1968; May 13, 1968. For similar Elavil ads, see *JAMA*, September 11, 1967; January 29, 1968; March 25, 1968. The ads posed ordinary-looking distressed patients under text such as, "Is it depression? He says 'It's my stomach' . . . but his other symp-

toms: functional somatic complaints, anxiety, insomnia, anorexia, feelings of guilt strongly suggest an underlying depression." Other examples: "She says 'I can't sleep'"; "He says 'I feel run down.'"

26. According to the *National Disease and Therapeutic Index*, vols. 9 (1962), 13 (1963/64), 17 (1964), and 21 (1965), and *National Disease and Therapeutic Index Review*, 1:1 (March 1970), 2:2 (December 1971), 4:1 (June 1973), and 7:2 (December 1976), diagnoses of mild or "neurotic" depression were approximately 4 million in 1962, out of 36 million "psychoneurotics"; this was fewer than the 7.4 million diagnoses of psychosis, and far fewer than the 12 million for straightforward "anxiety reaction." Diagnoses for mild depression surpassed those for psychosis in 1965. By 1968, depression diagnoses had reached nearly 8 million, while anxiety reactions had remained in the vicinity of 12 million. By 1971, both were at approximately 11 million; by 1975, depression, at 18 million diagnoses, had overtaken anxiety, with 13 million. See also Steven Secunda, *The Depressive Disorders: Special Report, 1973* (Rockville, MD: National Institute of Mental Health, 1973).

27. R. W. Shepherd, M.D., "Is Your Depression a Real Tiger?" *Vogue*, June 1978, 111.

28. John Schwab et al., "Sociocultural Aspects of Depression in Medical Inpatients: Frequency and Social Variables," *AMA Archives of General Psychiatry* 17 (1967): 533–38; Schwab et al., "Current Concepts of Depression: The Sociocultural," *International Journal of Social Psychiatry* 14 (1968): 226–34. For similar results, see also Eugene Levitt and Bernard Lubin, *Depression: Concepts, Controversies, and Some New Facts* (New York: Springer, 1975).

29. See, e.g., Triavil ad, *JAMA*, September 27, 1965; Elavil ad, *JAMA*, September 11, 1967; Etrafon ad, *American Journal of Psychiatry*, August 1966.

30. Etrafon ad, *JAMA*, March 11, 1968.

31. See, e.g., Elavil ad, *JAMA*, September 27, 1965; Tofranil ad, *American Journal of Psychiatry*, September 1968; Sinequan ad, *JAMA*, October 12, 1970; Deprol ad, *JAMA*, February 14, 1972; Elavil ad, *JAMA*, May 27, 1974, and September 22, 1975.

32. See, e.g., Ronald Fieve, *Moodswing: The Third Revolution in Psychiatry* (New York: Morrow, 1975); Huber, "The Blues and How To Chase Them"; Roisum, "How to Recognize Suicidal Depression"; "How to Beat the Blues," *Better Homes and Gardens*, February 1970, 116–17, 122; "McGovern's First Crisis," *Time*, August 7, 1972; "The Most Common Mental Disorder," *Time*, August 7, 1972, 16; "Depression and Electroshock," *Newsweek*, August 7, 1972, 20; "Coping with Depression," *Newsweek*, January 8, 1973, 51; Mercer and Sachar, "Complete Book of Depression."

33. Percy Knauth, *A Season in Hell* (New York: Harper and Row, 1975).

34. Myrna Weissman and Gerald Klerman, "Sex Differences and the Epidemiology of Depression," *AMA Archives of General Psychiatry* 34 (1977): 98–111; R. Hirschfeld and C. Cross, "Epidemiology of Affective Disorders: Psychosocial Risk Factors," *AMA Archives of General Psychiatry* 39 (1982): 35–46; D. Rice and J. Kepecs, "Patient Sex Differences and MMPI Changes, 1958–1969," *AMA Archives of General Psychiatry* 23 (1970): 185–92; S. Rosenthal, "Changes in a Population of Hospitalized Patients with Affective Disorders, 1945–1965," *American Journal of Psychiatry* 123 (1966): 671–81.

35. Nathan Kline, "No Fun? No Lust? Antidepressant May Bring New Life to Your Life,"*Vogue*, July 1975, 104–5. For representative examples, see Kosner, "What to Do When You're *Really* Depressed"; "Mercer and Sachar, "Complete Book of Depression"; Allison Robbins, "New Ups for Old Downs," *New York*, May 25, 1981; Dan Kaercher, "Depression: New Treatments for Our Number One Mental Health Problem," *Better Homes and Gardens*, April 1982, 23–28.

36. Triavil ad, *JAMA*, February 21, 1972; Sinequan ad, *JAMA*, October 12, 1970, and September 7, 1970; Vivactil ad, *JAMA*, April 13, 1970. See also advertisements in which a statue or puppet of a woman is returned to life and color, most often to enjoy time with her children. Vivactil ads, *JAMA*, January 1, 1968; March 18, 1968; September 2, 1968.

37. Tofranil ad,*American Journal of Psychiatry*, July 1968; Triavil ad, *JAMA*, front cover, January 6, 1975. See also, e.g., Elavil ad,*American Journal of Psychiatry*, October 1966; Tofranil ad, *American Journal of Psychiatry*, September 1967; Triavil ad, *JAMA*, January 6, 1975.

38. Helen De Rosis and Victoria Pellegrino, *The Book of Hope: How Women Can Overcome Depression* (New York: Bantam Books, 1976 [five printings from Macmillan 1976–77; five more from Bantam 1977–79]), 325 ("chronic, low-grade," living by "shoulds"); 36 ("rage, grief, and hopeless despair"), 335–36 ("dissatisfaction in the housewife role"; "right to be yourself"; skydiving). See also De Rosis interview in Kosner, "What to Do When You're *Really* Depressed," 285, and Mercer and Sachar, "Complete Book of Depression," 91; Wolfe, "When the Blues Don't Go Away," 132. De Rosis used Karen Horney as her theorist. For more on the fascinating history of psychoanalysis in the development of feminist thought, see Mari Jo Buhle, *Feminism and Its Discontents: A Century of Struggle with Psychoanalysis* (Cambridge: Harvard University Press, 1998). For a very different approach focusing on marginalized women, see Phyllis Chesler's worldwide bestseller *Women and Madness* (Garden City, NY: Doubleday, 1972), which argued that as a form of madness depression was actually a sane choice for desperate women, since "it is safer for women to become 'depressed' than physically violent," especially in mental institutions (41–46).

39. Maggie Scarf, *Unfinished Business: Pressure Points in the Lives of*

Women (Garden City, NY: Doubleday, 1980); Scarf, "Women Depressed—Why?" *Vogue*, August 1980, 255, 283–84; Scarf, "Women and Depression: It's the Price They Pay for Feeling," *New Republic*, July 5 and 12, 1980, 25–29; Scarf, "Why Women Are Depressed," *Newsweek*, September 8, 1980, 81–82. For an earlier example of similar logic, see Knox, "Blues Really Get You Down."

40. Weissman and Paykel, *Depressed Woman*, 217, 211–12.

41. Intriguing evidence of class dimensions of the illness can be found in the appearance in the 1990s of a wholly unprecedented figure in popular culture: the black woman depressive. As described in the pages of middle-class magazines like *Ebony* and *Essence* or in a handful of popular-format books, black women depressives were middle class or upwardly mobile. When an institution for upwardly mobile African Americans, the sorority Delta Sigma Theta, launched a campaign to help black women combat depression in the late 1990s, it seemed to confirm the notion that nervous illnesses came with social advancement. In addition to its public-health benefits, then, attentiveness to depression among black women could serve as a claim to medical citizenship in affluent America. The politics of depression among black women, however, differed in telling ways from the politics of nervous illnesses as articulated by white feminists from Charlotte Perkins Gilman to Betty Friedan. Portraits of depression in forums such as *Ebony* took greater pains than their white-culture counterparts to draw connections between professionals' suffering and the "blues" that threatened all African Americans because of the additional burden of racism and (often) poverty. "African-Americans have been systematically dehumanized by the social structure," the magazine quoted a psychology professor as saying in 1995; "African Americans more than any other group in this culture should be most depressed due to racism." Just as Gilman, Friedan, De Rosis, and others had appropriated for white women the existential angst that had long ennobled elite men, some black feminists sought to broaden that narrative to include the psychological impact of racism and economic hardship. It was a cultural gesture of middle-class status that left open the possibility of much broader inclusion of African Americans. Meri Nana-Ama Danquah, *Willow Weep for Me* (New York: Norton, 1998); Angela Mitchell with Kinnese Herring, *What the Blues Is All About* (New York: Berkeley, 1998). See also Julia Boyd, *Can I Get a Witness: For Sisters When the Blues Are More Than a Song* (New York: Dutton, 1998). For the Delta Sigma Theta campaign, see Kelly Starling, "Black Women and the Blues: Why So Many Sisters Are Mad and Sad," *Ebony*, May 1999, 140–44. See also Marie Saunders, "Depression: The Blues Within," *Essence*, April 1980, 91, 134–39; Lynn Norment, "Why Women Get Depressed," *Ebony*, April 1981, 84–89; "How to Beat Depression: Experts Say You Can," *Ebony*, August 1995, 102–6.

42. See Abigail Cheever, "Prozac Americans: Depression, Identity, and Self-

hood," *Twentieth Century Literature* 46 (Fall 2000): 346–69, for a similar argument following a trajectory from Walker Percy to William Styron. See also, e.g., Fieve, *Moodswing;* Knauth, *Season in Hell,* and William Styron, *Darkness Visible: A Memoir of Madness* (New York: Random House, 1990), both of which depict the authors as immobilized and incapable of work during their depressions; Wolfe, "When the Blues Don't Go Away"; Knox, "Blues Really Get You Down"; Kosner, "What to Do When You're *Really* Depressed"; Mercer and Sachar, "Complete Book of Depression"; Erica Goode, "Beating Depression," *U.S. News and World Report,* March 5, 1990, 48–56; Ayd, *Recognizing the Depressed Patient.* For an exception to this general rule, see Lesley Hazelton, *The Right to Feel Bad: Coming to Terms with Normal Depression* (Garden City, NY: Doubleday, 1984).

43. "Pills to Fight Mental Depression," *Business Week,* April 13, 1963, 76.

44. Galton, "Depression," 82.

45. "The Doctor Talks about Depression," *McCall's,* May 1958, 4, 151, 513; "Psychiatry: Injections for Depression," *Time,* May 7, 1965, 75.

46. Ayd, *Recognizing the Depressed Patient,* 133. Compare, e.g., the groundbreaking first report on Iproniazid, which examined the antidepressant's effect on "chronic" hospital patients mostly "in the schizophrenic classification," with the early reports by Borrus and Selling on Miltown (see chap. 1); H. P. Loomer, J. C. Saunders, and N. S. Kline, "A Clinical and Pharmaco-dynamic Evaluation of Iproniazid as a Psychic Energizer," *Psychiatric Research Reports* 8 (1957): 129–41; Emma Harrison, "TB Drug Is Tried in Mental Cases," *New York Times,* April 7, 1957, 86. See also the first high-profile publication about antidepressants, Linford Rees, "Treatment of Depression by Drugs and Other Means," *Nature,* April 9, 1960, 114–20, which held that "mild depressive states" were usually treated by "social, psychological and symptomatic measures," whereas "severe" depressions called for "electroplexy" (electroshock) or what the author considered to be the less effective antidepressants (119). See also Harold Himwich, "Psychoactive Drugs," *Postgraduate Medicine,* January 1965, 35–44: "Milder depressions, commonly seen in private practice, are managed in the same way as the neuroses [i.e., meprobamate]" (39, 43–44). Meanwhile, while waiting for imipramine to begin working, "the patient should be hospitalized to prevent suicide" (40).

47. Hutschnecker, "If You're Depressed." See also, e.g., "Tranquilizers and Other Psychoactive Drugs: How Well Do They Work?" *Consumer Reports,* October 1967, 547–48; Wolfe, "When the Blues Don't Go Away"; "Coping with Depression," *Newsweek,* January 8, 1973, 51.

48. One simple measure of this can be found by searching the *Reader's Guide to Periodical Literature* for "depression, mental" and for "antidepressants." The

first produces a slew of articles in the 1960s, 1970s, 1980s, and, to a much lesser extent, the 1990s. The second produces almost no references before the 1990s but a vast number thereafter.

49. The side effects included sweating, rapid heartbeat, blurred vision, sedation, dry mouth, and, for the users of one major class of the antidepressants (MAOIs), the possibility of death if a special diet was not carefully followed. For 1970s-era summaries of antidepressants' side effects, see, e.g., Leo Hollister, "Clinical Use of Psychotherapeutic Drugs II: Antidepressant and Antianxiety Drugs and Special Problems in the Use of Psychotherapeutic Drugs," *Drugs* 4 (1972): 383–86; Hollister, "Drug Therapy," *New England Journal of Medicine*, June 1, 1972, 1195–99.

50. Absolute numbers rose from an estimated 9.5 million in 1964 to 23.1 million in 1971 to approximately 30 million in 1975, where the figure remained until the appearance of Prozac in 1988. For full figures, and sources, see appendix B.

51. Hugh Parry et al.,"National Patterns of Psychotherapeutic Drug Use," *AMA Archives of General Psychiatry*, June 1973, 769. Psychiatrists prescribed 30% of all antidepressants but only 13% of antianxiety drugs; general practitioners prescribed 40–50% of all antianxiety drugs, sedatives, and amphetamines, but only 37% of all antidepressants—similar to their 34% rate for antipsychotics. Antidepressants were prescribed for mental illness 62% of the time (antipsychotics 69%), antianxiety drugs only 38% of the time. As late as 1978 one study found that psychiatrists knew more about antidepressants than general practitioners, but that the reverse was true for antianxiety medications. See Richard Gottlieb et al., "The Physician's Knowledge of Psychotropic Drugs: Preliminary Results,"*American Journal of Psychiatry*, January 1978, 29–32. Perhaps as a consequences of these patterns, one study suggested that antidepressants, like antipsychotics, may have been more prevalently prescribed for nonwhite and poor Americans, although the evidence is too thin to establish this with certainty. See Carl Chambers, "An Assessment of Drug Use in the General Population," in *Drug Use and Social Policy: An AMS Anthology*, ed. Jackwell Susman (New York: AMS, 1972), 50–123.

52. David Greenblatt, Richard Shader, and Jan Koch-Weser, "Psychotropic Drug Use in the Boston Area: A Report from the Boston Collaborative Drug Surveillance Program,"*AMA Archives of General Psychiatry*, April 1975, 518–21. Other studies found similar results for use in the past year. See, e.g., Glen Mellinger, Mitchell Balter, and Dean Manheimer, "Patterns of Psychotherapeutic Drug Use among Adults in San Francisco,"*AMA Archives of General Psychiatry*, November 1971, 384–94, which reported that nearly 13% of women and 8% of men had used a minor tranquilizer in the past year, whereas only 1.5% of either gender had used an antidepressant. Approximately 3% had used an antipsycho-

tic (388). Parry et al., "National Patterns of Psychotherapeutic Drug Use," replicated these findings at a national level: 15% of all respondents reported use of an antianxiety medicine in the past year, whereas only 2% reported antidepressant use; 1% reported antipsychotic use. A 1979 survey by the same group found these ratios still relatively unchanged: 8% of men and 14% of women had used an antianxiety agent in the past year, whereas only 1.5% of men and 2.8% of women had used an antidepressant. See Glen Mellinger and Mitchell Balter, "Prevalence and Patterns of Use of Psychotherapeutic Drugs: Results from a 1979 National Survey of American Adults," in *Epidemiological Impact of Psychotropic Drugs*, ed. C. Tognoni, C. Bellantuono, and M. Lader (New York: Elsevier/North-Holland Biomedical Press, 1981), 126.

53. Karl Rickels and Peter Hesbacher, "Psychopharmacologic Agents: Prescription Patterns of Non-Psychiatrists," *Psychosomatics*, December 1977, 37–40, found that of approximately 170 patients in seven family practices, 40 were diagnosed as "anxious," of whom 34 were given tranquilizers; 80 were diagnosed as having "mixed anxiety and depression," of whom 50 were given tranquilizers and only 13 antidepressants; only 21 were diagnosed as having depression by itself, and of these, 7 were given antianxiety medication and 10 were given antidepressants. See also David Raft et al., "Inpatient and Outpatient Patterns of Psychotropic Drug Prescribing by Nonpsychiatrist Physicians," *American Journal of Psychiatry*, December 1975, 1309–12, which found that in a general hospital, minor tranquilizers were "prescribed most often and with the least justification" while "antidepressants were given less often than would be justified."

54. Alan Schatzberg and Charles Nemeroff, eds., *The American Psychiatric Press Textbook of Psychopharmacology*, 2nd ed. (Washington, DC: American Psychiatric Press, 1998).

55. Healy, *Creation of Psychopharmacology*, 198–224.

56. Healy, *Let Them Eat Prozac*.

57. Healy, *Creation of Psychopharmacology*, 220–22.

58. Elizabeth Wilson, *Psychosomatic: Feminism and the Neurological Body* (Durham, NC: Duke University Press, 2004), 19. Intriguing support for this theory comes from psychopharmacologist Karl Rickels, mentioned in chapter 2 for his theory that wealthier patients responded best to the more expensive minor tranquilizers (he offered similar thoughts on true antidepressants versus Deprol). Rickels spent a fascinating chapter of his research career investigating what he called "non-specific factors" in drug treatment. The term referred to all elements of drug therapy not the direct result of pharmacological action, such as "suggestion; the doctor-patient relationship; the patient's personality, expectation, and attitudes; and innumerable other factors." In a double-blind study in which neither Rickels nor his subjects knew who received real drugs and who placebos,

for example, Rickels found that "compliant" subjects had a greater than normal rate of response to real drugs and a less than normal rate of response to placebos. His theory was that such patients were eager to "go along with the doctor" and "report improvement" but needed some—any—"active pharmacological effect" to work from. Another even more intriguing experiment found that if subjects were told they were receiving a sedative but were actually given either amphetamines or a placebo, they could not distinguish between small amounts of the active drug and the inert compound. If they were truthfully informed that the experiment involved a stimulant versus placebo, however, they had little difficulty drawing the distinction. As Rickels concluded, "The set of expectations produced by the experimenter . . . overcame the pharmacological effect of [the amphetamine]." See Karl Rickels, "Drugs in the Treatment of Neurotic Anxiety and Tension: Controlled Studies," in *Psychiatric Drugs: Proceedings of a Research Conference Held in Boston* [from the 1965 meeting of the APA's Regional Research Conference on Psychiatric Drugs], ed. Philip Solomon (New York: Grune and Stratton, 1966), 225–38. See also Rickels, ed., *Non-Specific Factors in Drug Therapy* [compiled from a symposium of the 4th World Congress of Psychiatry] (Springfield, IL: Charles C Thomas, 1968). For a much more in-depth examination of such reasoning, see Richard DeGrandpre, *The Cult of Pharmacology: How America Became the World's Most Troubled Drug Culture* (Durham, NC: Duke University Press, 2006).

59. Healy, *Creation of Psychopharmacology*; Elliott Valenstein, *Blaming the Brain: The Truth about Drugs and Mental Health* (New York: Free Press, 1988).

60. "Pills to Fight Mental Depression," *Business Week*; Galton, "Depression"; "New Ways to Treat Depression," *Good Housekeeping*, July 1975, 149; Kline, "No Fun? No Lust?"; Nathan Kline, *From Sad to Glad: Kline on Depression* (New York: Putnam, 1974); Fieve, *Moodswing*; Knauth, *Season in Hell.* See also, e.g., "Injections for Depression," *Time*, May 7, 1965, 75; Wolfe, "When the Blues Don't Go Away"; "Coping with Depression," *Newsweek*, January 8, 1973, 51; "What You Should Know about Mental Depression," *U.S. News and World Report*, September 9, 1974, 37–40; "Hormones for Depression," *Newsweek*, January 6, 1975, 56; Seymour Kety, "It's Not All in Your Head," *Saturday Review*, February 21, 1967; Kosner, "What To Do When You're *Really* Depressed;" Mercer and Sachar, "Complete Book of Depression"; Robbins, "New Ups for Old Downs"; Kaercher, "Depression: New Treatments for Our Number One Mental Health Problem." Maggie Scarf's *Unfinished Business* also devoted several pages to describing the neurochemical basis of depression and its treatments (250–67).

61. William Laurence, "Drug Called a 'Psychic Energizer' Found Useful in Treating Mental Illness," *New York Times*, December 22, 1957, 97. See also Nathan Kline's psychopharmacological utopianism in "Drugged Future?" *Time*,

February 24, 1958, 35–37 (among other things he speculated about developing a drug to heighten "extrasensory perception"); Robert DeRopp, *Drugs and the Mind* (New York: Grove Press, 1961), x (Kline quotation), 7 ("chemical processes"), 14 ("cruelty of the tyrant"), 282 ("chemopsychiatric era"); Kline in *Psychotropic Drugs in the Year 2000: Use by Normal Humans* (Springfield, IL: Charles C Thomas, 1971), 78–85.

62. Solomon Snyder, *Brainstorming: The Science and Politics of Opiate Research* (Cambridge: Harvard University Press, 1989).

63. David Leff, "Brain Chemistry May Influence Feelings, Behavior," *Smithsonian*, June 1978, 64–70; Rita Christopher, "The Mystery of Moods—It's All in the Mind," *Maclean's*, October 1, 1979, 46–49; Charles Panati, "Brain Breakthroughs: Your Body's Own Drugs for Pleasure and Pain," *Futurist*, February 1980, 21–26. For other popular coverage of enkephalins and endorphins, see, e.g., M. Clark and M. MacPherson, "The Brain's Own Opiate: Enkephalin," *Newsweek*, February 23, 1976, 79+; "Blood Bath: Isolation of Leu-Endorphin by Frank Ervin and Roberta Palmour," *Time*, November 21, 1977, 119; S. Drake, "Curing the Mind: Beta-Endorphin," *Newsweek*, August 29, 1977, 68–69; Richard Restak, "Brain Makes Its Own Narcotics!" *Saturday Review*, March 5, 1977, 6–11; "Brain's Opiate: Beta-Endorphin," *Time*, January 23, 1978, 98; Barbara Villet, "Opiates of the Mind," *Atlantic Monthly*, June 1978, 82–89; T. Cohen and R. Kushner, "Endorphins: Miracle or Mirage?" *Harper's Bazaar*, March 1978, 137+; "Painkillers: Lasker Awards," *Time*, December 4, 1978, 96; Jack Finsher, "Natural Opiates in the Brain," *Human Behavior*, January 1979, 28–32; Ellen Switzer, "Body's Own Controllers Soothe Anxiety, Stop Pain," *Vogue*, November 1979, 361+; Toby Cohen, "Beyond Valium," *New York*, February 5, 1979; Tom Alexander, "Body Telling the Mind: Endorphin Researcher C. Pert," *Fortune*, September 8, 1980, 97; "Chemical Reaction May Be the Villain: Beta-Endorphins and Obesity," *USA Today*, April 1981, 11; Joseph Carey, "The Brain Yields Its Secrets to Research," *U.S. News and World Report*, June 3, 1985, 64–65.

64. Seymour Rosenblatt with Reynold Dodson, *Beyond Valium: The Brave New World of Psychochemistry* (New York: G. P. Putnam's Sons, 1981), 237–43.

65. Richard Restak, *The Mind* (New York: Bantam Books, 1988), written to accompany the PBS special of the same name; Anthony Smith, *The Mind* (New York: Viking, 1984); Michael Gazzaniga, *Mind Matters: How Mind and Brain Interact to Create Our Conscious Lives* (Boston: Houghton Mifflin, 1988), quotation from 179; Paul H. Wender and Donald F. Klein, *Mind, Mood and Medicine: A Guide to the New Biopsychiatry* (New York: Farrar, Straus, and Giroux, 1981), quotation from 197; Richard Restak, *The Brain: The Last Frontier* (New York: Doubleday, 1979); Nancy Andreasen, *The Broken Brain: The Biological Revolution in Psychiatry* (New York: Harper and Row, 1984); Floyd E. Bloom and

Arlyne Lazerson, 2nd ed. *Brain, Mind, and Behavior* (New York: Freeman, 1988); Robert Ornstein and Richard F. Thompson, *The Amazing Brain* (Boston: Houghton Mifflin, 1984).

66. Jon Franklin, *Molecules of the Mind* (New York: Atheneum, 1987), 94 (human condition "up for grabs"), 217 *("zzzt! zzzt!"),* 277 ("intelligence boosters"), 91 ("the potentiality for at least two diseases").

67. Tranxene ad, *JAMA*, March 20, 1978; February 23, 1979; initial euphoria, January 21, 1983; drug-seeking behavior, March 19, 1982.

68. Kline, "No Fun? No Lust?" See also, e.g., Michael Halberstam, "Common, Insidious, Mysterious, *New York Times,* July 30, 1972, E1.

69. See, e.g., Walter Sullivan, "Depletion of Hormone Linked to Depression," *New York Times,* August 25, 1982; Margot Slade and Katherine Roberts, "Ideas and Trends: A Chemical Key to Depression," *New York Times,* July 29, 1984.

70. Robbins, "New Ups for Old Downs"; Laurence Cherry, "The Good News about Depression," *New York,* June 2, 1986, 33–40; Franklin, *Molecules of the Mind,* 118.

71. "Drugs Most Frequently Prescribed in Physicians' Offices," *World Almanac and Book of Facts,* 2000 and 2002; M. Olfson and G. L. Klerman, "The Treatment of Depression: Prescribing Practices of Primary Care Physicians and Psychiatrists," *Journal of Family Practice,* December 1992, 627–35; Olfson and Klerman, "Trends in the Prescription of Antidepressants by Office-Based Psychiatrists," *American Journal of Psychiatry,* April 1993, 571–77; Olfson et al., "Antidepressant Prescribing Practices of Outpatient Psychiatrists," *AMA Archives of General Psychiatry,* April 1998, 310–16; Olfson et al., "National Trends in the Outpatient Treatment of Depression," *JAMA,* January 9, 2002, 203–9; J. M. Zito et al., "Rising Prevalence of Antidepressants among U.S. Youths," *Pediatrics,* May 2002, 721–27; M. E. Hemels, G. Koren, and T. R. Einarson, "Increased Use of Antidepressants in Canada: 1981–2000," *Annals of Pharmacotherapy,* September 2002, 1375–79.

72. Kramer, *Listening to Prozac.* Reviews included David Gates, "The Case of Dr. Strangedrug," *Newsweek,* June 14, 1993, 71; Anastasia Toufexis, "The Personality Pill," *Time,* October 11, 1993, 61–62; Walter Reich, "What the Medicine Said," *Wilson Quarterly,* Autumn 1993, 74–77; Jacob Sullum, "Beyond Blue," *Reason,* December 1993, 45–47; David Rothman, "Shiny Happy People," *New Republic,* February 14, 1994, 34–38; John Stapert, "Curing an Illness or Transforming the Self?" *Christian Century,* July 13–20, 1994, 684–86; Daniel Freedman, "On Beyond Wellness," *New York Times Book Review,* August 8, 1993, 6–9; Susan Brink, "No More Ms. Meek," *U.S. News and World Report,* November 8, 1993; Reid Cushman, "You've Got Personality," *Nation,* November 8, 1993, 537–40; Avery Brown and John Loengard, "Miracle Worker," *People,* November 15,

1993, 153–56; Charles Medawar, "Through the Doors of Deception," *Nature*, March 24, 1993, 369–71; Sherwin Nuland, "The Pill of Pills," *New York Review of Books*, June 9, 1994, 4–8. Memoirs: Elizabeth Wurtzel, *Prozac Nation* (New York: Riverhead, 1995), and Lauren Slater, *Prozac Diary* (New York: Random House, 1998). A *Reader's Guide to Periodical Literature* search for "Prozac" finds approximately 50 articles published in the five years before Kramer's book and approximately 150 in the next five years.

73. Sara Rimer, "With Millions Taking Prozac, a Legal Drug Culture Arises," *New York Times*, December 13, 1993; Geoffrey Cowley, "The Culture of Prozac," *Newsweek*, February 7, 1994, 41.

74. For popular-media coverage of celebrity depressives, see, e.g., Goode, "Beating Depression," 50, or Rosie O'Donnell, "To Tell the Truth: In Millions of Families, Depression Is an Illness You Just Don't Talk About. Rosie Breaks the Silence," *Rosie*, September 2001, 112–43 (it is in *Rosie* that the highlighted celebrities' stories appear opposite ads for prescription antidepressants). In addition to Wurtzel's and Slater's memoirs, see, e.g., Kitty Dukakis, *Now You Know* (New York: Simon and Schuster, 1990), and Styron, *Darkness Visible*.

75. L. J. Simoni-Wastila, "Gender and Psychotropic Drug Use," *Medical Care*, January 1998, 88–94; A. A. Hohmann, "Gender Bias in Psychotropic Drug Prescribing in Primary Care," *Medical Care*, May 1989, 478–90; M. Olfson et al., "Antidepressant Prescribing Practices of Outpatient Psychiatrists," *AMA Archives of General Psychiatry*, April 1998, 310–16; J. A. Sirey et al., "Predictors of Antidepressant Prescription and Early Use among Depressed Outpatients," *American Journal of Psychiatry*, May 1999, 690–96; C. A. Melfi et al., "Racial Variation in Antidepressant Treatment in a Medicaid Population," *Journal of Clinical Psychiatry*, January 2000, 16–21; D. G. Blazer et al., "Marked Differences in Antidepressant Use by Race in an Elderly Community Sample: 1986–1996," *American Journal of Psychiatry*, July 2000, 1089–94; C. M. Roe et al., "Gender- and Age-Related Prescription Drug Use Patterns," *Annals of Pharmacotherapy*, January 2002, 30–39.

76. Mimi Bluestone, "Fighting Depression with One of the Brain's Own Drugs," *Business Week*, February 22, 1988, 156–57.

77. G. Cowley and K. Springen, "The Promise of Prozac," *Newsweek*, March 26, 1990, 38–42.

78. J. Mendes, "Products of the Year," *Fortune*, December 17, 1990, 72–78. Prozac starred alongside Rollerblades and Microsoft's Windows 3.0.

79. Greg Critser, "Oh, How Happy We Will Be: Pills, Paradise, and the Profits of the Drug Companies," *Harper's*, June 1996, 39–48. For a representative sampling of many other examples, see "The Boom in Depression," *Economist*, August 8, 1992, 73–75; Duff McDonald, "Smile with Prozac—and Laugh to the Bank

with Eli Lilly," *Money,* April 1996, 88–89; Michael Lemonick, "Beyond Depression," *Time,* May 17, 1999; Susan Headden,"The Big Pill Push," *U.S. News and World Report,* September 1, 1997, 67–68+; Thomas Moore, "Hard to Swallow," *Washingtonian,* December 1997, 68–71+.

80. For a recent examination of direct-to-consumer ads, see Julie Marie Donohue, "Pharmaceutical Promotion in an Age of Consumerism" (Ph.D. diss., Harvard University, October 2003).

81. This particular phrasing comes from Kramer's cover story in *Psychology Today,* July–August 1993, 42–53.

82. Natalie Angier, "Drug Works, but Questions Remain," *New York Times,* December 13, 1993; Fran Schumer, "Bye-Bye, Blues: A New Wonder Drug for Depression," *New York,* December 18, 1989, 48; Tracy Thompson, "Seeking the Wizards of Prozac," *Saturday Evening Post,* March–April 1994, 50–53+. See also, e.g., Thompson, "Coming Back with Prozac," *Pittsburgh Post-Gazette,* November 28, 1993, and Thompson, "Shadow on the Brain: One Woman's Search for the Roots of Chronic Depression,"*Washington Post,* July 9, 1995. For a different perspective on this aspect of antidepressants, see Jonathan Metzl, *Prozac on the Couch: Prescribing Gender in the Era of Wonder Drugs* (Durham, NC: Duke University Press, 2003).

83. Cowley and Springen, "Promise of Prozac"; Toufexis, "Personality Pill"; Michael Lemonick, "The Mood Molecule" [cover story], *Time,* September 29, 1997, 74–82; Walter Kirn, "Living the Pharmaceutical Life," *Time,* September 29, 1997, 82; Geoffrey Cowley and Anne Underwood, "A Little Help from Serotonin," *Newsweek,* December 29, 1997–January 5, 1998, 78–81. See also, e.g., Bluestone, "Fighting Depression"; Schumer, "Bye-Bye, Blues"; Natalie Angier, "Moody News: Can a Pill Called Prozac End Depression?" *Mademoiselle,* April 1990, 229, 263; Philip Elmer-Dewitt, "Depression: The Growing Role of Drug Therapies," *Time,* July 6, 1992, 56–60; Angier, "Drug Works, but Questions Remain"; Thompson, "Seeking the Wizards of Prozac"; "Does Prozac Stifle Creativity?" *USA Today Magazine,* April 12, 1995; Clark Barshinger, "The Gospel According to Prozac," *Christianity Today,* August 14, 1995, 34–38; Patricia Lopez Baden, "Beyond the Blues," *Better Homes and Gardens,* September 1995, 56; Samuel Barondes, *Better Than Prozac: Creating the Next Generation of Psychiatric Drugs* (Oxford: Oxford University Press, 2003).

84. Elmer-Dewitt, "Depression"; Robert Wright, "The Coverage of Happiness," *New Republic,* March 14, 1994, 24; John Stapert, "Curing Illness or Transforming the Self? The Power of Prozac," *Christian Century,* July 13–20, 1994, 686; Sharon Begley and Debra Rosenberg, "One Pill Makes You Larger, and One Pill Makes You Small," *Newsweek,* February 7, 1994, 36–41; Carla Koehl, "Happier Days Ahead? The Future of Mind Drugs," *Newsweek,* April 21, 1997. See also,

e.g., Cowley and Springen, "Promise of Prozac"; Gates, "Case of Dr. Strangedrug"; Toufexis, "Personality Pill"; Lisa Krieger, "Key to Pharmaceutical Mood Control: Chemical Has Role in Mental Health," *Albany Times Union*, April 1, 1994, C1; Judith Hooper and Hannah Bloch, "Targeting the Brain: The 3-Lb. Organ that Rules the Body Is Finally Giving up Its Secrets; Goodbye, Oedipus," *Time*, Fall 1996 Special Issue, 14; Daniel Goleman, "Coming Soon to a Prozac Nation," *New York Times*, November 24, 1996; Daniel Goleman, "Research on Brain Leads to Pursuit of Designer Drugs," *New York Times*, November 19, 1996; Lemonick, "Mood Molecule."

85. Toufexis, "Personality Pill," 61.

86. A. A. Hohmann, "Gender Bias in Psychotropic Drug Prescribing in Primary Care," *Medical Care*, May 1989, 478–90; L. Simoni-Wastila, "Gender and Psychotropic Drug Use," *Medical Care*, January 1998, 88–94; Jonathan Metzl and Joni Angel, "Assessing the Impact of SSRI Antidepressants on Popular Notions of Women's Depressive Illnesses," *Social Science and Medicine* 58 (2004): 578.

87. Prozac ad, *Cosmopolitan*, September 1997. See also, e.g., references in Schumer, "Bye-Bye, Blues"; Cowley and Springen, "Promise of Prozac"; Angier, "Moody News"; Rimer, "With Millions Taking Prozac, a Legal Drug Culture Arises."

88. "No More Ms. Meek," *U.S. News and World Report*, November 8, 1993, 77.

89. Prozac ad, *American Journal of Psychiatry* 155 (1998): A7. For a different take on this and other advertisements emphasizing how they subtly promised a return of patriarchal control over Prozac-taking women, see Metzl, *Prozac on the Couch*, 152–55.

90. Federal expenditures rose from less than $1 billion in 1980 to over $8 billion in 1995, and state-level spending reached similar levels. Meanwhile, federal outlays for battling narcotics in Latin America tripled as well. By 1995, drug offenders had grown from a small fraction of prison inmates to an outright majority at the federal level and a substantial 20% in state and local prisons, even as the nation's total prison population more than doubled. See Ethan Nadelmann, "Drug Prohibition in the U.S.: Costs, Consequences, and Alternatives," in *Crack in America: Demon Drugs and Social Justice*, ed. Craig Reinarman and Harry G. Levine (Berkeley: University of California Press, 1997), 292–94, and Ted Carpenter, *Bad Neighbor Policy: Washington's Futile War on Drugs in Latin America* (New York: Palgrave Macmillan, 2003).

91. Racial disparities from Troy Duster, "Pattern, Purpose, and Race in the Drug War: The Crisis of Credibility in Criminal Justice," in *Crack in America*, ed. Reinarman and Levine, 260–87.

92. Curtis Marez, *Drug Wars: The Political Economy of Narcotics* (Minneapolis: University of Minnesota Press, 2004), esp. 245.

93. Nancy Campbell, "Regulating 'Maternal Instinct': Governing Mentalities of Late-Twentieth-Century U.S. Illicit Drug Policy," *Signs,* Summer 1999, 895–923.

94. For an unusual example (at least in the mass media), see Lynette Holloway, "Seeing a Link between Depression and Homelessness," *New York Times,* February 7, 1999, 3 (Week in Review section).

95. Sharon Lerner, "Chemical Reaction," *Ms.,* July–August 1997, 56–61. See also Margot Lyon, "C. Wright Mills Meets Prozac: The Relevance of 'Social Emotion' to the Sociology of Health and Illness," in *Health and the Sociology of Emotions,* ed. Veronica James and Jonathan Gabe (Oxford: Blackwell, 1996); Judith Gardiner, "Can Ms. Prozac Talk Back? Feminism, Drugs, and Social Constructionism," *Feminist Studies,* Fall 1995, 501–18; Betsy Morris, "Executive Women Confront Midlife Crisis," *Fortune,* September 18, 1995, 60–62+.

96. For a similar argument about Prozac memoirs, see Metzl, *Prozac on the Couch.*

97. Healy, *Let Them Eat Prozac;* Joseph Glenmullen, *Prozac Backlash: Overcoming the Dangers of Prozac, Zoloft, Paxil, and Other Antidepressants with Safe, Effective Alternatives* (New York: Simon and Schuster, 2000). See, e.g., Benedict Carey, "Panel to Debate Antidepressant Warnings," *New York Times,* December 13, 2006.

98. Among many examples, see Peter and Ginger Ross Breggin, *Talking Back to Prozac: What Doctors Won't Tell You about Today's Most Controversial Drug* (New York: St. Martin's, 1995); Peter Breggin, *The Antidepressant Fact Book: What Your Doctor Won't Tell You about Prozac, Zoloft, Paxil, Celexa, and Luvox* (Cambridge, MA: Perseus, 2001); Jay Cohen, *Over Dose: The Case against the Drug Companies: Prescription Drugs, Side Effects, and Your Health* (New York: Jeremy P. Tarcher/Putnam, 2001); Marcia Angell, *The Truth about Drug Companies: How They Deceive Us and What to Do about It* (New York: Random House, 2004). For a sampling of magazine coverage focusing on Prozac, see, e.g., G. Gowley, "A Prozac Backlash," *Newsweek,* April 1, 1991, 64–68; Alexander Cockburn, "Paradigms of Power: The Case of Eli Lilly," *Nation,* December 7, 1992, 690–92; Tanya Bibeau, "The Dark Side of Psychiatric Drugs," *USA Today Magazine,* May 1994, 44–48; Mark Nichols, "Questioning Prozac," *Macleans,* May 23, 1994, 36–41; Christine Gorman, "Prozac's Worst Enemy," *Time,* October 10, 1994, 64–66; Clyde Haberman, "Blaming Prozac: 90's Version of Twinkie Defense in Subway Bombing Trial," *New York Times,* February 6, 1996, B3; Erica Goode, "Once Again, Prozac Takes Center Stage, in Furor," *New York Times,*

July 18, 2000, F1; Oliver Baker, "Are Doctors Too Free with Prozac?" *New Scientist*, March 9, 2002, 15.

Conclusion: Better Living through Chemistry?

1. Richard DeGrandpre, *The Cult of Pharmacology: How America Became the World's Most Troubled Drug Culture* (Durham, NC: Duke University Press: 2006).

addiction: to barbiturates, 90–96, 133–37; defined pharmacologically, 113–19; and disease model and its critics, 97–101, 104–5, 106–13; and feminism, 137–49; and lines between "drugs" and "medicines," 83–105, 113–21, 125–37, 184–88, 194–99; to Miltown, 83–84; terminology, 87; to Valium, 113–21, 137–49

addiction treatment (medical specialty), 132, 138–39

advertising: of antidepressants, 154, 157–59, 178, 183, 185–86; direct-to-consumer, 41–44, 178–79; and gender, 68–73, 73–75, 80–82; of Librium, 29, 40–41; and masculinity, 66–73, 75–76; of Miltown, 27–29, 45–46, 58–59, 67–71, 80–81, 107–8; of prescription drugs, 23–29, 128, 197–98; of Prozac, 178, 183, 185–86; of Valium, 145–46

Advertising Age, 29

Advisory Commission on Narcotic and Drug Abuse, 98–99, 102

Alcoholics Anonymous (AA), 97, 98

alcoholism, 83, 87, 132

alcoholism movement, modern, 98

American Home Products, 25–29

American Medical Association (AMA), 19, 26, 31, 36, 40, 97, 111

American Psychiatric Association, 156–57

amphetamines, 9, 133–37, 154

anhedonia, 153–54

Anslinger, Harry, 88–90, 96, 97–98, 101

antidepressants: before Prozac, 154–55, 157–61, 164–69; and biological psychiatry, 166–75; discovery of, 20; and

epidemiology, 176–77; and gender, 181–91. *See also specific drug names*

antipsychiatry, 19

antipsychotics: definition of, 8; discovery of, 17–18

anxiety: biological ideas of, 68–72; and gender, 50–56, 62–73; and postwar culture, 51–56; psychoanalytic ideas of, 31–38

Atarax, 34–35, 42

Aventyl, 128, 157–58

Ayd, Frank, 19–20, 116–17, 155, 157, 165

barbiturates, 9, 90–96, 126–37

Beard, George, 49–50

Beck, Aaron, 155

Benzedrine, 154

benzodiazepines. *See* Librium; Valium

Berger, Frank, 21, 24, 114

biological psychiatry, 30–31; and antidepressants, 166–75, 179–82; and consumerism, 169–75; and gender, 68–73, 179–82; supported by pharmaceutical industry, 167–68. *See also* monoamine hypothesis

Boggs Act, 93–96

Boston Women's Health Book Collective, 130–31

Bureau of Drug Abuse Control, 105, 115

Bureau of Narcotics and Dangerous Drugs, 115

Carter Products, 21–29, 106, 107, 108, 114–15, 157

Chayet, Neil, 116–17

chlordiazepoxide. *See* Librium

chlorpromazine. *See* antipsychotics

cognitive therapy, 155

Cold War, 65

Cole, Jonathan, 117
congressional investigations or hearings: and Drug Abuse Control Act, 96–105; and establishment of Drug Enforcement Administration, 115–19; and Kefauver, 24, 41–42, 127–28; and Edward Kennedy, 139; and Valium addiction, 139, 144–47
consumerism: and antidepressants, 169–75, 177–82; and Prozac, 177–82; and tranquilizers, 23, 43–44, 60–62, 140
Consumers Union, 133
cosmetic psychopharmacology. *See* Kramer, Peter
counterculture, 131–32
Cowen, Belita, 131
Charpentier, Paul, 18
crack cocaine, 184, 187

depression: before Prozac, 152–64; and biological psychiatry, 166–68; and gender, 159–63
Deprol, 157
de Ropp, Roberg, 170
De Rosis, Helen, 161–62
detail men, 25
Dexamyl, 75–77
Diagnostic and Statistical Manual, 32–33, 153
diazepam. *See* Valium
Didion, Joan, 136–37
Dodd, Thomas, 102–4
Drug Abuse Control Act (1965), 96–105
Drug Abuse Warning Network, 138
drug advertising. *See* advertising
Drug Enforcement Administration (DEA), 115, 138; Schedule of Controlled Substances, 86, 116–19
drug regulation, 22, 88–96, 101–5, 113–19
drug war. *See* war on drugs

Eagleton, Thomas, 156–57
Eddy, Nathan, 92–93, 95–96, 99–101, 108, 113
Elavil, 154, 157–58
endorphins and enkephalins, 171, 174
epidemiology: of antidepressants, 165, 176–77; of anxiety, 33, 51–54; of depression, 158–59; of tranquilizer use, 38, 39, 73, 128–29, 138
Equanil, 25–29. *See also* Miltown
Essig, Carl, 92, 108, 112–13
Etrafon, 159
Eutonyl, 164

Federal Bureau of Narcotics, 88–90
feminism: and depression, 161–62; and minor tranquilizers, 73–82; opponents of, 162–63; and Prozac, 179–89, 200–201; and Valium addiction, 123–49
Fieve, Ronald, 169
fluoxetine hydrochloride. *See* Prozac
Food and Drug Administration (FDA), 22, 35, 88, 91, 94, 101–8, 113, 167, 178, 190
Ford, Betty, 122, 141–44
Franklin, Jon, 172–75
Freud, Sigmund. *See* psychoanalysis
Friedan, Betty, 75–82, 139

Gazzaniga, Michael, 172
Geigy Pharmaceuticals, 20, 29
gender: and addiction, 137–49; and advertising, 68–73, 73–75, 80–82; and antidepressants, 181–91; and anxiety, 50–56, 62–73; and biological psychiatry, 68–73, 179–82; and depression, 159–63; and nervous illnesses, 50–56; and tranquilizers, 73–82, 137–49; and women's health movement, 130–31, 138–40. *See also* feminism; men; women
Glenmullen, Joseph, 190
Gordon, Barbara, 122, 141–44

Harris, Isbell, 92–96, 99–101
Harrison Anti-Narcotic Act, 88
Health Research Group, 128
Healy, David, 20, 154, 167–68, 190
Hollingshead, August, 51–52
Hollister, Leo, 109, 114
homemaking. *See* mothers
Hoyt, Henry, 21, 24–25
Humphrey-Durham Amendment, 91, 126
Huxley, Aldous, 65

imipramine, 20
interpersonal therapy, 155–56
iproniazid, 20, 170

Kefauver, Estes, 24, 41–42, 127–28
Kennedy, Edward, 139
Kennedy, John F., 98
Klerman, Gerald, 155–56
Kline, Nathan, 19–20, 34, 154, 155, 160, 168–69, 170, 174
Knauth, Percy, 160, 169
Kolb, Lawrence, 89, 112
Kramer, Peter, 146, 150–52, 175–82
Kuhn, Roland, 20, 154

Laborit, Henry, 17–18
Lemere, Frederick, 83–84, 108, 110
Librium, 8–9, 29, 39–41. *See also* Valium
Listening to Prozac (Kramer), 150–52, 175–82

Maginnis, Cynthia, 144–47
Mailer, Norman, 54–55
major tranquilizers. *See* antipsychotics
Marplan, 154
masculinity: and addiction to tranquilizers, 110–11; and Miltown phenomenon, 62–73
May, Rollo, 52–53
Mead, Margaret, 53, 159
media. *See* popular media
Medical and Pharmaceutical Information Bureau, 43, 164
Medical Letter on Drugs and Therapeutics, 39
men: and Miltown, 62–73, 75–76; and nervous illnesses, 49–50, 53–55. *See also* gender; masculinity
meprobamate. *See* Equanil; Miltown
Merck Sharp and Dohme, 154, 157
methadone, 97
Miltown: addictiveness of, 83, 105–13; as antidepressant, 154; commercial development of, 24–29; as consumer good, 59–62; decline of, 38–39; discovery of, 21; and masculinity, 62–73; and medical practice, 31, 34–38; as middle-class

drug, 57–59, 61; as minor tranquilizer, 8–9
minor tranquilizers. *See specific drug names*
Mintz, Morton, 127–28
monoamine hypothesis (of depression), 166–68; and Prozac, 179–81. *See also* biological psychiatry
monoamine oxidase inhibitors (MAOIs), 166
Morris, Desmond, 72
mothers, 45, 52–53, 140, 160–63
Myserson, Abraham, 153–54

Nader, Ralph, 128
National Institute of Mental Health, 93, 95, 117, 128–29
National Institute on Alcohol Abuse and Alcoholism, 132
National Institute on Drug Abuse, 132
National Women's Health Network, 131
nervous illnesses: and anxiety, 49–56; and class, 49–56; and depression, 176–77; and gender, 50–56
neurasthenia, 49–51
Niamid, 154
Nixon, Richard, 133
Nysewander, Marie, 97, 99, 138–39

patient demand, 44–46
patients: depicted as addicts, 133–48; depicted as losing masculinity, 62–73; depictions of, during Prozac phenomenon, 176–82, 184; organizing drug treatment groups, 144–48; and patients' rights movement, 129–30; seeking control over own drug therapy, 44–46; as targets of drug advertising, 41–44
patients' rights movement, 129–30
Paxil, 1–3
penicillin, 22
Pert, Candace, 172–73
Pfizer, 154
pharmaceutical industry: investigations into practices of, 127–28; postwar boom of, 21–29, 193, 196–97; resisting drug regulation, 94–95, 102–5, 106–7, 112, 115, 119

physicians: depicted as "pushers," 137–
47; psychiatrists, postwar lack of, 33–
34; role of, in Miltown phenomenon,
35–38, 192–93
Physicians Desk Reference, 26–27, 112,
136, 137–38
popular media: and antidepressants,
168–69, 176, 177–89; and biological
psychiatry, 171–75; and depression,
152–53, 156–57, 165; and feminism,
179–82, 184; and prescription drug
addiction, 122–49; and Miltown phenomenon, 42–43, 53–54, 60–62, 63–
66
Prozac, 150–52, 175–91; and biological
psychiatry, 179–82; and consumerism,
177–82; and feminism, 181–91; as
response to Valium crisis, 181–91
Prozac Diary (Slater), 176
Prozac Nation (Wurtzel), 176
psychoanalysis: and concepts of depression, 153–54; and Miltown, 15, 30–38
psychosomatic medicine, 32–33

Reagan, Nancy, 148
Reagan, Ronald, 184
Redlich, Frederick, 51–52
Reisman, David, 54–55
Restak, Richard, 172
Ritalin, 128–29, 158
Robinson v. California, 98
Roche Pharmaceuticals, 26, 104, 106,
109, 115–16, 119
Rolling Stones, the, 131–32
Rosenblatt, Seymour, 171–72

Scarf, Maggie, 162–63
Schedule of Controlled Substances, 86,
116–19
Schlafly, Phyllis, 162
Schlesinger, Arthur, 55, 66
Seaman, Barbara, 131
selective serotonin reuptake inhibitors
(SSRIs). *See* Prozac
Selye, Hans, 30–31
Serax, 80
Serentil, 128
serotonin, 179–82
Sinequan, 75, 160

Slater, Lauren, 176
Smith Kline and French, 18, 25–26, 154
sociobiology. *See* biological psychiatry
Striatran, 68
Styron, William, 176
"supermoms," 181–83
Susann, Jacquelyn, 136–37
Szasz, Thomas, 19

Ted Bates and Company, 27–28
Temperance, 87
therapy: cognitive-behavioral, 155;
interpersonal, 155–56. *See also*
psychoanalysis
Thorazine. *See* antipsychotics
Tofranil, 161
tranquilizers. *See specific drug names*
Tranxene, 174
Triavil, 157–58, 160–61
tricyclics, 166

utilization studies. *See* epidemiology

Valium: and addiction scare, 137–49;
addictiveness of, debated, 109–19;
as antidepressant, 154; decline of,
145–46; discovery and commercial
development of, 39–41; as minor
tranquilizer, 8–9; Prozac as response
to crisis of, 181–91
Valley of the Dolls (Susann), 136–37
Vivactil, 75, 78, 160

Wallace Pharmaceuticals. *See* Carter
Products
war on drugs: origins of, 87–90; and
Prozac, 184, 187–88; and Valium
addiction, 122–49
Weissman, Myrna, 155–56, 162–63
William Douglas McAdams, Inc., 29
Wilson, Edward, 72
Wolfe, Sidney, 128
women: and advertising, 68–73, 73–75,
80–82; and depression, 159–63; and
Prozac, 181–91; and tranquilizers in
postwar era, 45, 73–82; and Valium
addiction, 137–49; and war against
drugs, 122–49. *See also* feminism;
gender

women's health movement, 130–31, 138–40

Women-Together, Inc., 144–47

wonder drugs, postwar discoveries of, 17

World Health Organization (WHO), Expert Committee on Addiction-Producing Drugs, 93, 99–101, 119

World War II, 22, 32

Wurtzel, Elizabeth, 176

Xanax, 123